Time in Early Modern Islam

The prophet Muhammad and the early Islamic community radically redefined the concept of time that they had inherited from earlier religions' beliefs and practices. This new temporal system, based on a lunar calendar and era, was complex and required sophistication and accuracy. From the ninth to the sixteenth centuries, it was the Muslim astronomers of the Ottoman, Safavid, and Mughal empires, and not those of Europe, who were responsible for the major advances in mathematics, astronomy, and astrology. Stephen P. Blake's fascinating study compares the Islamic concept of time, and its historical and cultural significance, across these three great empires. Each empire, while mindful of earlier models, created a new temporal system, fashioning a new solar calendar and era and a new round of rituals and ceremonies from the cultural resources at hand. The hysteria that accompanied the end of the first Islamic millennium in 1591 also created a unique collection of apocalyptic prophets and movements in each empire. This book contributes not only to our understanding of the Muslim temporal system, but also to our appreciation of the influence of Islamic science on the Western world.

Dr. Stephen P. Blake is Professor Emeritus at St. Olaf College, Northfield, Minnesota, and Senior Research Fellow, Center for Early Modern History, University of Minnesota. His books include *Shahjahanabad: The Sovereign City in Mughal India, 1639–1739* (Cambridge University Press 2002), and *Half the World: The Social Architecture of Safavid Isfahan, 1590–1722* (1999).

To Meg, with love as always

To my children – Andrew, Edward, John, Paul, and Rachel

Time in Early Modern Islam

Calendar, Ceremony, and Chronology in the Safavid, Mughal, and Ottoman Empires

STEPHEN P. BLAKE
Center for Early Modern History, University of Minnesota

CAMBRIDGE
UNIVERSITY PRESS

CAMBRIDGE UNIVERSITY PRESS
Cambridge, New York, Melbourne, Madrid, Cape Town,
Singapore, São Paulo, Delhi, Mexico City

Cambridge University Press
32 Avenue of the Americas, New York, NY 10013-2473, USA

www.cambridge.org
Information on this title: www.cambridge.org/9781107030237

First published 2013

Printed in the United States of America

A catalog record for this publication is available from the British Library.

Library of Congress Cataloging in Publication Data
Blake, Stephen P.
Time in early modern Islam : calendar, ceremony, and chronology in the
Safavid, Mughal, and Ottoman empires / by Stephen P. Blake.
p. cm.
Includes bibliographical references and index.
ISBN 978-1-107-03023-7 (hardback)
1. Islamic calendar. 2. Astronomy – Religious aspects – Islam. 3. Time – Religious aspects – Islam. 4. Islam and science – History. 5. Iran – History – Safavid dynasty, 1501–1736. 6. Mogul Empire. 7. Turkey – History – Ottoman Empire, 1288–1918. I. Title.
CE59.B63 2013
529′.327–dc23 2012023174

ISBN 978-1-107-03023-7 Hardback

Contents

Preface

As the last of the great world religions, Islam entered a crowded spiritual arena. Although Mecca, the home of the prophet Muhammad, contained the central shrine of the pagan Bedouins, it also housed sizable Jewish and Christian communities. As a result, an early aim of the new community was to distinguish itself from the beliefs and practices of its neighbors. And one of its most radical departures was a thoroughgoing redefinition of the concept of time. In the Quran Allah had ordained a new calendar. The year was to be strictly lunar, 354 days, divided into twelve months of twenty-nine and a half days each. Although many early states had adopted lunar calendars (the Babylonians, Egyptians, Jews, and Greeks, for example), their eras were lunisolar. That is, in order to keep the calendar (and its ceremonies and rituals) in rough synchronization with the seasons, these states added an extra month every three years or so. However, because Allah had expressly prohibited a thirteenth month, the strictly lunar era of the Muslims contained no lengthened years. Rather, the 354-day year of the lunar era regressed against the 365-day year of the solar era at the rate of 11 days per cycle. In thus cutting the calendar loose from the seasons, the new religion gave birth to an intense preoccupation with time – defining it, measuring it, celebrating its peak moments. In the years after Muhammad's death the introduction of new rituals accelerated the process of redefinition. A month of fasting and a pilgrimage month required a precise determination of the phases of the moon, and the five daily prayers demanded an accurate timetable of the sun's movement across the daytime sky.

Because in the premodern world time was measured by the movements of the heavenly bodies (the earth, sun, moon, the five visible planets, and the visible stars) and not by mechanical devices (clocks or watches), the new temporal system created a new dedication to astronomy. In premodern Eurasia, astronomy was the queen of the sciences, reaching (according to a modern historian) "a much higher level of accuracy and sophistication than any other science."[1] And from the middle of the ninth century until the middle of the sixteenth it was the Islamic astronomers in the observatories of Isfahan, Maragha, and Samarqand, not those in Paris, London, or Rome, who achieved the major advances in the science. Not until Tycho Brahe in the late sixteenth century did the Europeans match the equipment, precision, and mathematical sophistication of the Islamic astronomers.

In the premodern Islamic world, the astronomer/astrologer (Persian, *munajjim*; Turkish, *muneccim*) was the time expert. Not only was he responsible for establishing the observational and mathematical parameters of the new temporal system, but he was also given the task of working out its astrological components. Although astrology also defined a temporal order dependent on the heavenly bodies, it was a completely unscientific construct. The time of the astrologer was symbolic, eschatological, and millennial. Having constructed a horoscope or chart of the heavens, the astrologer foretold the future for individuals, communities, and nations. He determined the appropriate time for undertaking or avoiding activities. He revealed the hidden patterns of the cosmos – the eclipses and conjunctions, the comets and shooting stars – that predicted the triumphs and disasters of the terrestrial world.

By the early modern period, three great empires had come to rule the Islamic world – the Mughal (1526–1739) in India; the Safavid (1501–1722) in Iran; and the Ottoman (ca. 1289–1923) in Anatolia, the Balkans, Syria, Egypt, the Arabian Peninsula, and North Africa. Although all three had inherited the basic Islamic concept, their different cultural environments (Indic for the Mughals, Zoroastrian for the Safavids, and Christian for the Ottomans) meant that in each state different temporal systems emerged. These systems varied in three ways – calendrically, ceremonially, and chronologically. The calendrical difference revolved around the addition of one or more solar calendars to the liturgical lunar, the necessity in an agrarian state for a calendar that tracked the seasons. The ceremonial difference, by contrast, centered on the adoption of new public rituals, celebrations that reflected both the dynasty's heritage and its need for legitimacy. Finally, the chronological issue was twofold: First, in an agrarian state with multiple

[1] Steven Weinberg, "The Missions of Astronomy," *New York Review of Books* 56 (22 Oct. 2009): 19.

eras, how were the several dating systems managed? How many eras were employed and how were documents, orders, and histories dated? Second, how did the three empires handle the millenarian hysteria of the late sixteenth century? The Grand Conjunction of Jupiter and Saturn in 1583 and the end of the first Islamic millennium in 1591 caused a great deal of political, social, and economic disruption, throwing up an unstable collection of prophets and sects.

Acknowledgments

I have long been interested in a comparative study of the three early modern Islamic empires. My introduction to the topic was in the early 1990s: two conferences at Harvard University on the political economies of the Mughal, Safavid, and Ottoman Empires. At that time I benefited from conversations with Bert Fragner, Metin Kunt, David Ludden, Burton Stein, and John Richards. My journey led me from Mughal India to Safavid Iran, and finally to the Ottoman empire. Along the way I received advice and encouragement from Stephen Dale, Andrew Newman, Charles Melville, Rudi Matthee, Muzaffar Alam, Chris Bayly, and Peter Bang. Two other early modern historians have been especially influential. Sanjay Subrahmanyam has written extensively and perceptively on a variety of topics in early modern Eurasian history, and Cornell Fleischer has been creative and especially insightful in his studies of millenarianism in the early modern Mediterranean world. Francis Robinson's article on shared knowledge was an early inspiration.

This book has been a long time coming, and I owe a great deal to the support of my family. I dedicate the final product to my children – Andrew, Edward, John, Paul, and Rachel – and to my wife, Meg. They provided entertainment and encouragement during the long months of composition.

Note on Transliteration

I have adopted a simplified system of transliteration. Arabic and Persian words are rendered according to the system of the *International Journal of Middle East Studies* but without the diacriticals. Ottoman Turkish words are given in their modern Turkish form, and common words follow their spelling in the *IJMES* Word List. For most dates I employ the Common Era but, because of my topic, I also provide a good number of Hijra Era equivalents.

Introduction

> What then is time? I know well enough what it is, provided that nobody asks me; but if I am asked what it is and try to explain, I am baffled.[1]

St. Augustine's (354–430) question – How to conceptualize something so ordinary and transparent as time? – has bedeviled scientists, philosophers, and social scientists from his era to ours. In his recent and immensely popular *A Brief History of Time*,[2] Stephen Hawking spelled out the revolutionary implications of Albert Einstein's (1879–1955) theory of relativity. In contrast to Isaac Newton (1643–1727), who considered time an objective aspect of the natural world, an absolute background unchanged by motion or matter, Einstein proposed a new relativistic model in which clocks ran slower or faster depending on their speed or location. The sociologist Emile Durkheim (1858–1917), a rough contemporary of Einstein, offered a parallel theory of temporal relativity – an argument about time from the social rather than from the cosmological perspective. Following Kant, Durkheim contended that the basic categories of the understanding (including, of course, time) were not given a priori but were social constructs.[3] Edmund Leach, the anthropologist, wrote, "We talk of measuring time, as if it were a concrete thing to be measured, but in fact we create time by

[1] Augustine, *Confessions and Inchiridon*, ed. and trans. Albert Cook Outler (Dallas: Library of Christian Classics, 1955), book 11, ch. 14.

[2] Stephen Hawking, *A Brief History of Time* (New York: Bantam Dell Publishing Company, 1988).

[3] Norbet Elias, *Time: An Essay*, trans. Edmund Jephcott (Oxford: Blackwell, 1992), 4. For an interesting overview of the various sociocultural approaches to time, see Peter Burke, "Reflections on the Cultural History of Time," *Viator: Medieval and Renaissance Studies* 35 (2004): 617–26.

creating intervals in social life. Until we have done so there is no time to be measured."[4] The relationships between things and activities are described in terms of time (and, of course, space) but time itself is not another economic resource to be conserved or wasted.

Although there are many ways of approaching the topic of time, the social or cultural point of view is the one taken here. From this perspective, there were three dimensions to Islam's radical restructuring of the temporal system: the calendrical, the ceremonial, and the chronological. The rest of this chapter briefly describes the origin and early development of the Islamic system – from its beginnings under the prophet Muhammad (ca. 570–632) to its final form at the end of the Abbasid empire (750–1258) – with a special emphasis on the radical nature of the new concept. Setting the stage for what follows, this introduction provides for each chapter a background, against which the characteristic contour of the individual temporal dimension – calendrical, ceremonial, or chronological – can be identified.

CALENDRICAL

In premodern societies, the most common units of time – the day, the month, and the year – were defined by the three great natural timekeepers – the earth, the moon, and the sun.

The rotation of the earth on its axis established the first natural division – light and dark – and the twenty-four-hour period was the shortest unit of the natural clock. Most societies further divided the daylight hours by the apparent movement of the sun across the sky. Because the axis of the earth was tilted at 23.5 degrees, the amount of light in a twenty-four-hour period varied with the seasons (except at the equator). Only at the equinoxes (equal nights) – the Vernal (21 March) and the Autumnal (21 September) – were day and night exactly twelve hours each.[5]

The second natural timekeeper, the moon, revolved around the earth and cycled through its phases in approximately 29.5 days. The interesting thing about the moon was the way its shape changed. When directly between the earth and the sun, it was invisible; as it orbited, however, it

[4] Janet Hoskins, *The Play of Time: Kodi Perspectives on Calendars, History, and Exchange* (Berkeley: University of California Press, 1993), 373. A comprehensive look at the various methodologies employed by anthropologists can be found in Nancy D. Munn, "The Cultural Anthropology of Time: A Critical Essay," *The Annual Review of Anthropology* 21 (1992): 93–123.

[5] Beulah Tannebaum and Myra Stillman, *Understanding Time: The Science of Clocks and Calendars* (New York: Whittlesey House, 1958), 93–4.

waxed, became full, and then waned to a sliver, disappearing entirely. The cycle was repeated again and again. Because these changes – from sliver, to half, to full, and back again – were so readily apparent and since the approximately thirty-day period mimicked other natural intervals – the female menstrual period and the cyclic behavior of certain marine creatures – the cycle assumed great importance in the ancient world and was the basis of many early calendars.[6]

The third natural timekeeper was the sun. Although it took the earth approximately 365.25 days to travel around the sun, astronomers determined the precise length of the solar year in two slightly different ways. The tropical solar year, employed by the astronomers of the Islamic tradition, was the time it took the sun, in its apparent movement, to return to the same reference point on the ecliptic – 365 days, 5 hours, 48 minutes, and 46 seconds. That reference point was typically the Vernal Equinox, where the celestial equator intersected the ecliptic (the plane of the earth's orbit around the sun). In astrological terminology it was the first point of Aries and was usually dated 21 March (although it could sometimes fall on the 19th or the 20th). The sidereal solar year, by contrast, employed in the Indic astronomical tradition, measured the apparent motion of the sun with reference to a fixed background star. The two definitions differed slightly on the length of the solar year – the sidereal being about twenty minutes longer than the tropical.

The word "calendar" comes from the Latin "calendarium," an interest register or account book. At its most basic, the calendar is a way of keeping track of the first natural time division – the day.[7] It is an abstract method of naming the days by allocating each to a week, a month, and a year. A guide to day-to-day activities, the calendar enables a society to fix its important rituals and festivals. It offers a way of recording and arranging the events of the past as well as of calculating commitments for the future. To produce such an abstract temporal system was the impetus behind most early attempts to observe and record the positions of the heavenly bodies.[8]

The Muslim day began at sundown. Although daily prayers were mentioned in the Quran, it is generally agreed that neither their number nor their timing had been established by the end of the prophet's life.

[6] Anthony Aveni, *Empires of Time: Calendar, Clocks, and Cultures* (New York: Basic Books, 1989), 86–7.

[7] *Encyclopedia Britannica*, 15th ed., s.v. "Calendar."

[8] Sir Harold Spencer Jones, "The Calendar," in C. Singer, ed., *History of Technology*, vol. 3 (London: OUP, 1957), 558.

Seventy-five to a hundred years later, however, during the eighth century, all this had changed. Five prayers had become standard but their exact time had not been specified: (1) *Salat al-maghrib* (sunset prayer), when the sun had disappeared over the horizon; (2) *Salat al-isha* (evening prayer), after twilight; (3) *Salat al-fadjr* (morning prayer), at daybreak; (4) *Salat al-zuhr* (noon prayer), after noon, the sun just beginning to decline; (5) *Salat al-asr* (afternoon prayer), the sun still high, white, and pure.[9]

The call to prayer by the muezzin (caller) was given five times a day:

> Prayer is better than sleep. God is Greatest, God is Greatest. I assert that there is no god but God. I assert that Muhammad is the Messenger of God. Come to the prayer. Come to salvation. God is Greatest, God is Greatest. There is no god but God.

Believers could begin their devotions at any time after hearing the call but, according to the law books, earlier was better.

The prayers were of unequal length – sunset had three *rakas* (bowings), evening four, daybreak two, noon four, and afternoon four. The noon prayer on Friday (the day of assembly) was special. The faithful gathered in a congregational mosque, and the usual noon prayer was shortened to two bowings. There was also a *khutba* (address). Not a Christian sermon, interpreting a passage from the sacred book, its form was fixed: Praise of God and the Prophet, a recitation from the Quran, an admonition of piety, and finally an invocation of God's blessing on the local political leader.

Because of the liturgical division of the day, Islamic astronomer/astrologers developed an early interest in the science of time keeping (*ilm al-miqat*). In the earliest centuries the times of prayer were defined by the sun: its appearance and disappearance and the length of its shadow. In the earliest astronomical handbooks (*zij*), therefore, there were extensive tables listing shadow lengths for various cities – for example, Baghdad, Cairo, and Damascus. Often as well, the handbooks included projections for nearby towns and cities.[10]

Because, however, the most important prayer of the week – the noon prayer on Friday – was scheduled for a particular moment rather than for a range of hours, an interest in more exact methods of time reckoning soon developed. By the early thirteenth century, *a muwaqqit* (timekeeper) had begun to appear on the staff of many mosques. At about the same

[9] *Encyclopaedia of Islam*, 2d. ed., s.v. "Miqat"; *Encyclopaedia of Islam*, 2d. ed., s.v. "Tarikh," 259.

[10] *Encyclopaedia of Islam*, 2d. ed., s.v. "Miqat."

time the name of another time specialist (the *miqati*), a *munajjim* who specialized in spherical astronomy and astronomical timekeeping but who was not a mosque official, became increasingly common in the astronomical/astrological literature. A man from Cairo, for example, compiled a set of tables displaying prayer times as a function of solar altitude and longitude while another put together a more comprehensive table – the time could be calculated for all latitudes from either the sun or the stars.[11]

With this commitment to more precisely dividing the day came a corresponding interest in more accurate time keeping devices. The *gnomon* (a vertical bar that casts a shadow) was the earliest instrument for telling time by the sun. But it was crude and very inaccurate and was soon replaced by the sundial. In the early thirteenth century a Cairo *miqati* wrote an extensive treatise on sundial theory and on the construction of sundials in mosques.[12] In 1276 the astronomer/astrologer Abu al-Abbas Ahmad ibn Umar al Sufi authored a treatise on the defects of sundials – their causes and correction.[13]

A further advance in the science of timekeeping was signaled by the appearance of the *clepsydras* (water clock). Invented in Babylonia and Egypt, the earliest specimen was discovered in the Temple of Karnak (fifteenth century BCE).[14] Although the sundial was unreliable, dependent on the season or the weather, the water clock enabled the *muwaqqit* to determine prayer times more exactly – without resort to astronomical tables or to the appearance of the sun or stars. The first Islamic water clocks were constructed in the early ninth century. Harun al-Rashid (r. 766–809), the legendary Abbasid ruler of Thousand-and-One-Nights fame, sent an elaborate time piece to the Holy Roman Emperor Charlemagne (r. 800–814) in 807. A sensation at the Frankish court, it was "... a marvelous mechanical contraption, in which the course of the twelve hours moved according to a water clock, with as many bronze little balls, which fell down on the hour ... there were also twelve horsemen who at the end of each hour stepped out of twelve open windows ..."[15] In 1205 CE the *munajjim* al-Rizwan completed a manuscript on astronomical clocks, with drawings illustrating their

11 Ibid.

12 Ibid.

13 Abdus Sattar Siddiqi, "Construction of Clocks and Islamic Civilization," *Islamic Culture* 1 (1927): 245–51.

14 *The International Encyclopedia of the Social Sciences*, s.v. "Time." 16:30.

15 David Ewing Duncan, *Calendar: Humanity's Epic Struggle to Determine a True and Accurate Year* (New York: Avon Books, 1998), 117.

mechanism and functioning. He also included a description of the famous clock in the great mosque of Damascus. It had been constructed by his father, Muhammad ibn Ali ibn Rustam al-Saati (the Horologist), in the third quarter of the twelfth century.[16] By the early fifteenth century, the water clock had spread across the Islamic world and had become the instrument of choice for the *muwaqqit* and the muezzin. This interest in more accurate timekeeping was reflected in the new name of the discipline – *ilm-i muwaqqit* (science of fixed times).[17]

Unlike the day, month, or year, the week was not determined by the movements of the earth, moon, or sun. The seven-day week of the Judeo-Christian world probably derived from the Babylonian and Egyptian belief that the seven heavenly bodies ruled the days:[18] Saturn governed Saturday, the Sun Sunday, the Moon Monday, Mars Tuesday, Mercury Wednesday, Jupiter Thursday, and Venus Friday. This method of dividing the month into seven-day segments was adopted first by the Jews and later by the Christians.[19]

Although Muhammad and the early Muslims accepted the seven-day Judeo-Christian week, the problem of finding a special day of worship was difficult. According to an early tradition, the prophet stated:

> The Jews have every seventh day a day, when they get together [for prayer] and so do Christians: therefore, let us do the same.[20]

The early Muslims chose Friday (sundown Thursday until sundown Friday) as their peak day, keeping it near but separate from the holy days of the other two religions. Perhaps, however, feeling the need to further differentiate themselves, they did not make Friday a day of rest. Unlike Saturday for Jews or Sunday for Christians, Friday for Muslims was a day of ordinary activity (except for the noon prayer). To further distinguish themselves, they also settled on different names for their days. Thus, Friday, the day of the special community prayer, was the "Day of Assembly" (Yaum al-Jama) and Saturday, under Jewish influence, was the Sabbath (Yaum al-Sabt). The other days, however, were simply the

[16] Siddiqi, "Construction of Clocks," 245–51.

[17] E. S. Kennedy, "Al-Biruni on the Muslim Times of Prayer," in E. S. Kennedy, *Studies in the Islamic Exact Sciences* (Beirut: American University of Beirut, 1983), 299–310.

[18] *Encyclopedia Britannica*, 15th ed., s.v. "Calendar"; Duncan, *Calendar*, 53–55.

[19] Eviatar Zerubavel, *The Seven Day Circle: The History and Meaning of the Week* (New York: The Free Press, 1985), 14–17.

[20] Ibid., 26.

first day (Sunday), the second day (Monday), the third day (Tuesday), the fourth day (Wednesday), and the fifth day (Thursday).[21]

In the early centuries of the first millennium, the inhabitants of the Arabian Peninsula had a strictly lunar calendar. The months were defined by the phases of the moon and were approximately 29.5 days each. The lunar year, comprised of twelve lunations, was divided into two parts: four months of peace (three centered on the pilgrimage month) when raiding and fighting were prohibited, and eight months in which warfare was allowed. However, because the lunar year contained about 354 days and the solar year of four seasons about 365, the pilgrimage month, which had originally been in the autumn, regressed against the seasons, making it progressively more difficult to find provisions for traveling and animals for sacrifice. As a result, in 412 the Arabs adopted a lunisolar calendar, intercalating a month every three years, placing it between Zu al-Hijja (the month of pilgrimage) and Muharram (the first month of the year).[22]

In 631, however, according to an obscure passage in the Quran, the prophet Muhammad was commanded to reform the pagan lunisolar calendar and era that he had inherited. "The number of months in the sight of Allah is twelve – so ordained by Him the day he created the heavens and the earth.... Verily *nasi* [the intercalation of a month] is an addition to unbelief: The Unbelievers are led to wrong thereby: ..."[23] This prohibition was later repeated by the prophet: "Oh People, the unbelievers indulge in tampering with the calendar in order to make permissible that which Allah forbade, and to forbid that which Allah has made permissible. With Allah the months are twelve in number ..."[24] The Quranic verse has been difficult to interpret, the meaning of *nasi* uncertain. Some eighth- and ninth-century authorities argued that it referred to the official in charge of the calendar, while others maintained that it meant the addition of an extra month. Although most Islamic astronomers agreed with the latter interpretation, it is worth noting that in the first centuries of the present era the *nasi* was the spiritual leader of the Jewish community. He was responsible, among other things, for determining the first day of the month (on the appearance of the new

[21] Nachum Dershowitz and Edward M. Reingold, *Calendrical Calculations* (Cambridge: Cambridge University Press, 1997), 63.

[22] *Encyclopaedia of Islam*, 2d. ed., s.v. "Tarikh."

[23] Quran 9: 36–7.

[24] *Encyclopaedia of Islam*, 2d. ed., s.v. "Tarikh."

moon) and for deciding when to intercalate an extra month. By the fourth century, however, the years of intercalation had become fixed, and the *nasi* no longer made these decisions ad hoc.[25]

In the first centuries after the death of Muhammad, the beginning of the month and the number of its days varied. A new month could not be declared until the first slim crescent had appeared and predicting this event was a major motivation behind the early Muslim interest in astronomy. Soon, however, in order to simplify astronomical calculations and to establish specific dates for rituals and celebrations, Islamic astronomers adopted a schematic calendar in which the months were given a definite number of days: (1) Muharram, thirty days; (2) Safar, twenty nine days; (3) Rabi I, thirty days; (4) Rabi II, twenty nine days; (5) Jumada I, thirty days; (6) Jumada II; twenty nine days; (7) Rajab, thirty days; (8) Shaban, twenty-nine days; (9) Ramadan, thirty days; (10) Shawwal, twenty-nine days; (11) Zu al-Qada, thirty days; and (12) Zu al-Hijja, twenty-nine or thirty days. The extra day was sometimes necessary because twelve revolutions of the moon totaled about 354.25 days. In a thirty-year cycle, the additional day was added in the second, fifth, seventh, tenth, thirteenth, sixteenth, eighteenth, twenty-first, twenty-fourth, twenty-sixth, and twenty-ninth years.[26] The names of the months were pre-Islamic and did not change but they soon lost their seasonal connotations. For example, Ramadan had originally meant "hot [summer]," Safar "yellow [fall]," Rabi I "the grazing season [spring]," and Jumada I "hard, frozen [winter]." In the prophet's sermon the months were further divided: the first, seventh, eleventh, and twelfth were holy.

CEREMONIAL

The new calendar featured a number of new ceremonies. In the first years after Muhammad there were only two festivals – Id al-Fitr and Id al-Qurban. Id al-Fitr, celebrating the end of the month-long Ramadan fast, lasted for three days – from the first to the third of Shawwal. Although exceptions were made for the sick, the young, the old, and the pregnant, the expectation was that no believer would eat or drink

[25] Andre Nehr, "The View of Time in Jewish Culture," in UNESCO, *At the Crossroads of Cultures and Time* (Paris: UNESCO, 1976), 163–5.

[26] Frank Parise, ed., *The Book of Calendars* (New York: Facts on File, 1982), 71.

from sunrise to sunset. Ramadan was a month of atonement and forgiveness; during this period the gates of heaven were said to be open and the gates of hell closed.[27]

The other major feast of the Islamic year was Id al-Qurban, the Festival of Sacrifice. It occurred during the pilgrimage month (Zu al-Hijja) and ran from the tenth to the thirteenth. One of the five pillars of Islam, the pilgrimage to Mecca was an obligation which only a minority of believers could fulfill. Mecca was the birthplace of the Prophet and the home of the Kaba, the great stone that, according to tradition, had been erected by Abraham. The most important action performed by the pilgrim was the sacrifice of an animal – a chicken, sheep, goat, cow, or camel. This sacrifice, on the tenth of the month, reenacted the near-sacrifice by Abraham of Ishmael, the son of Hagar and, performed by Muslims everywhere, symbolized the solidarity of the worldwide community.[28]

A third celebration, which developed four or five centuries later, was the prophet's birthday. Because Muhammad's actual date of birth was unknown, the day of his death (12 Rabi I) was chosen as the occasion for the festivities. The celebration seems to have begun in the late twelfth century. Soon after, rulers and nobles began to mount more elaborate commemorations, which came to include recitations from the Quran as well as stories from the prophet's life – in verse or in a combination of verse and prose.[29]

The festival of Ashura, celebrated during the first ten days of Muharram, commemorated the death of the Imam Husain (son of Ali and grandson of the prophet) at Kerbala on 10 October 680. During the first nine days the celebrants donned mourning clothes, ate simply, and listened to stories of Husain's sufferings. Groups of half-naked men paraded through the streets flagellating themselves, crying, and moaning. On the tenth day, a replica of Husain's coffin was publicly displayed and his funeral was reenacted. Although the ritual was most popular in Shiite communities, the tale of Husain's passion and death had the power to touch the lives of Muslims of whatever persuasion.[30]

[27] G. E. von Grunebaum, *Muhammadan Festivals* (New York: Henry Schuman, 1951), 51–3, 56–65.
[28] Ibid., 15–36.
[29] Ibid., 73–7.
[30] Ibid., 85–9.

CHRONOLOGICAL

From the chronological perspective the study of time is the study of the year. In order to locate significant religious, cultural, and political events many early states organized their years into eras. An era was a numbered collection of years reckoned from a specific date. The founding date, known as the epoch, typically marked an important event: the founding of a dynasty or the birth of a prophet or king. By introducing eras, societies were able to traverse time in two directions – earlier events could be related one to the other, sooner or later, and future events could be predicted and located with accuracy.[31]

During the decade and a half following the Prophet Muhammad's migration (*hijra*) from Mecca to Medina his followers gave the years names rather than numbers. The second was the Year of Permission, the fifth was Congratulations on Marriage, and the year of his death was Farewell. However, Umar (634–44), the second caliph, soon realized that in his rapidly expanding community a more conventional chronology was needed. Thus in 638 he established the Hijra Era. The departure of Muhammad from Mecca in 622 was chosen as the starting point because, according to tradition, the prophet's followers could not agree on the date of his birth. The inaugural day of the new era, however, was not the actual date of the prophet's emigration but was rather the first day of the lunar year in which it took place. Thus, 1 Muharram AH 1 was 16 July 622. The first documented example of the new era was an Egyptian papyrus of 22 AH (642–43 CE).[32]

Munajjim (Astronomer/Astrologer)

In medieval and early modern Islam time was defined by the movements of the seven heavenly bodies – sun, moon, Mercury, Mars, Venus, Jupiter, and Saturn. And as it was the responsibility of the *munajjim* to chart and interpret their movements, he became the time specialist, the man in charge of the Islamic temporal system in all its phases – calendrical, ceremonial, and chronological. According to the early encyclopedist, Jabbir b. Hayyan, the *munajjim* "... must have a mastery of astronomy ... [which is] a

[31] Paul Ricoeur, *Time and Narrative*, trans. Kathleen Blamey and David Pellauer (Chicago: University of Chicago Press, 1988), 71–2.

[32] E. G. Richards, *Mapping Time: The Calendar and Its History* (Oxford: Oxford University Press, 1999), 234.

description of the sky and what it contains [and a mastery of] astrology [which is] the gift of the planets."[33]

In his role as calendrical expert the *munajjim* charted the movements of the sun and the moon, deciding on the proper times for the daily prayers and determining when the Ramadan fast began and ended. As the ceremonial expert, he established dates for the important festivals and rituals in all the relevant calendars. He also labeled the days – auspicious or inauspicious depending on the person involved and the activity contemplated. And if the day were auspicious, he chose the exact moment to initiate the activity or event. Finally, as chronological expert the *munajjim* identified the first day of the year in the various calendars and eras and calculated, for historians, accountants, and recordkeepers, equivalent dates across the several temporal systems. He was also the repository of historical knowledge, able to provide the names and dates of battles, rulers, dynasties, and prophets. In addition, he was often called on to predict and interpret important astronomical events – equinoxes, solstices, eclipses, comets, shooting stars, and planetary conjunctions.

How was the *munajjim* trained? How did he acquire his knowledge and skills? Three paths were possible. The first was self-instruction: He could pick up some rudimentary astronomy and mathematics, study an astrological text (such as Ptolemy's *Tetrabiblios*), and learn to read an almanac, cast a horoscope, and work an astrolabe. A second, although relatively untrodden, path was formal study at a madrasa (college) attached to an observatory. There, a talented young man could acquire the skills of his profession in an academic setting. The most common path by far, however, was to become an apprentice, to learn at the feet of a successful master.[34]

Having finished his training, the newly minted *munajjim* needed three pieces of equipment: an astrolabe, an almanac, and a dust board. The astrolabe, an instrument consisting of a disc and a pointer, was used for both observation and computing – to determine the positions of the heavenly bodies and to calculate the time of day. The almanac contained, among other things, a series of tables listing the positions of the seven heavenly bodies for each day of the year. Finally, the dust board was used for mathematical calculations, to translate the general information from

[33] *Encyclopaedia of Islam*, 2d. ed., s. v. "munadjdjim."

[34] George Saliba, "The Role of the Astrologer in Medieval Islamic Society," *Bulletin d'Etudes Orientales* 44 (1992): 45–68.

the astrolabe and almanac into specific predictions for individuals – rulers, soldiers, merchants, or peasants.[35]

Islamic astronomy/astrology drew its inspiration from a variety of sources. Although the pre-Islamic Arabs of the peninsula had a basic knowledge of the stars and planets, the major astronomical and astrological influences on the early Muslim scientists were Greek: Ptolemy, and to a lesser extent, Aristotle.[36] Claudius Ptolemy (ca. 100–70), the famous astronomer, mathematician, astrologer, and geographer, was an Egyptian who lived in Alexandria. His great treatise, which remained the center of Islamic astronomy for centuries, was translated into Arabic in ninth-century Baghdad as *The Almagest* (The Greatest). The original Greek title, *The Mathematical Compilation*, became in time *The Greatest Compilation*, and, befitting its position in Islamic astronomy, finally ended up simply as *The Greatest.* Its listing of 1,022 stars in 48 constellations was the basic framework for Islamic and European astronomy until the seventeenth century. For Ptolemy, the earth was the center of the universe – the sun, moon, and the five planets revolving around it. He was also responsible for the standard work on astrology, *The Tetrabiblos.*[37]

Although the Greek tradition was preeminent, there were other influences as well. The early Islamic scientists had been exposed to Babylonian theories and discoveries through the works of Iranian astronomers. The most important Pahlavi astronomical treatise, *Zik-i Shatro-Ayar*, was translated into Arabic in ca. 790 as *Zij al-Shah.* And the early Islamic *munajjim*, Abu Mashar (787–886), wrote that astrology was the teaching of the Iranians.[38] The Indians also played a role. A major Indian text – Brahmagupta's *Brahma-Sphuta Siddhanta* (ca. 628) – was translated into Arabic in circa 770 under the title *Zij al-Sindhind.* Indian astronomy, however, had also been influenced by the Babylonians and the Greeks, and so an exact determination of the various influences on Islamic astronomy/astrology is difficult to make.[39]

Although some early modern Islamic *munajjims* worked primarily as astronomers – observing the heavens and making mathematical calculations – the vast majority earned their living as astrologers. Islamic

[35] Ibid.; Seyyed Hossein Nasr, *An Introduction to Islamic Cosmological Doctrines* (Cambridge, MA: Harvard University Press, 1964), 107–65; *Encyclopaedia of Islam*, 2d. ed., s.v. "al-Nudjum."

[36] *Encyclopaedia of Islam*, 2d. ed., s.v. "al-Nudjum."

[37] John David North and Roy Porter, *The Norton History of Astronomy and Cosmology* (New York: Norton, 1994), 107–21.

[38] Ibid., 175–6.

[39] Ibid., 163–4.

astrology, based primarily on the work of Ptolemy, assumed an intimate connection between celestial patterns and terrestrial events. It was predicated on the idea that the earth rested motionless at the center of the universe and was the pivot of the eight orbits that encircled it – those of the seven heavenly bodies plus an eighth comprised of the fixed stars. The task of the astrologer was to chart the movements of the heavenly bodies and to predict and interpret their influence on earthly institutions and individuals.[40]

The principal tool of the Islamic astrologer was the horoscope. Invented by the Greeks in the first century BCE, the word meant "a look at the hours." As drawn up by the *munajjim* (using an astrolabe, almanac, and dust board), the horoscope was a diagram of the heavens – a map of the sky at a particular time and place, exactly locating the sun, moon, planets, and stars. In order to more easily analyze and interpret the heavenly chart, the early Greeks divided the horizon into twelve houses or signs – each bearing one of the familiar names, Aries, Sagittarius, Taurus, or Aquarius. Each horoscope was unique and, analyzed and interpreted by the astrologer, yielded a specific prediction about the future.

Islamic astrology was divided into four parts. The first was general astrology and applied to nations, regions, or religions. The predictions were based on a horoscope drawn up for the first day of the year (usually the Vernal Equinox). They were natural – the conditions of the crops and the occurrence of floods, droughts, or plagues; political – the succession of prophets or rulers and the outcome of wars; or historical – the impact of celestial events like planetary conjunctions or eclipses on states, regions, and religions.

The second was genethlialogy, predicting the significant events in the lives of individuals – marriage, disease, or death – from their birth charts (horoscopes at birth).

The third was elections or choices. Determining (by the means of a horoscope) the best time to undertake an activity – when to have intercourse, consult a doctor, elect a new king, or embark on a war. It also involved deciding which days were auspicious or inauspicious. According to al-Jahiz (ca. 781–869):

> . . . the astrologers had examined the days of the week, judging them and appraising them in the interests of the king. They said, "Each day has its star which dominates it and its character which this star necessarily confers upon it.

[40] *Encyclopaedia of Islam*, 2d. ed., s.v. "ilm al-haya."

Accordingly, they determined for each day of the week the tasks appropriate to it. The auspicious and inauspicious character of the days of the week depended on the planets to which they were dedicated. Similarly, the hours of the day were dedicated to the seven planets.[41]

The fourth aspect of Islamic astrology was interrogations. Here the issue was not when but whether. Should I marry this woman or attack this enemy? The horoscope was cast at the time the question was asked.[42]

Because of the radical redefinition of time by the early Muslim community, the observatory as an institution was an Islamic creation. Ptolemy had, of course, made observations, but his instruments were small, and he was not part of any larger scientific body nor was he given any royal or state patronage. After Ptolemy, in the centuries before Muhammad, the situation remained basically unchanged.

By the ninth century, however, Islamic *munajjims* had become more interested in accuracy. They began to correct, refine, and extend the tables in Ptolemy's *Almagest*. Fresh observations of the sun and moon, calculating the latitude of the newly conquered towns and cities in order to establish the *qibla* (direction of prayer), and determining more precisely the location and movements of the five planets – all of these required new instruments of observation. New instruments, in turn, entailed a new form of organization. They were much larger, no longer portable, and had to be installed in a permanent, specially designed complex. A staff of some size was necessary, the array of equipment and tasks beyond the capacity of a single individual. And they required a significant investment – both for their purchase and installation and for their maintenance. Until at least the mid-sixteenth century, therefore, the greatest advances in both astronomy and astrology took place in the Islamic world. Until that point, European astronomy and astrology consisted primarily of Latin translations from the Arabic and Persian works of Islamic scientists.

OBSERVATORY AT ISFAHAN

The first example of a functioning observatory was during the reign of the Seljuq ruler Jalal al-Din Malik Shah (r. 1072–92). Because most scholars think that the Seljuq court was peripatetic, the principal mystery surrounding the observatory is its location. Arguments for Nishapur and Rey have been

41 *Encyclopaedia of Islam*, 2d. ed., s.v. "al-Nudjum."
42 Ibid.

advanced but Isfahan, Malik Shah's favorite residence, seems the most likely possibility. At first, the ruler wanted a complete revision of Ptolemy, but when his court *munajjims* pointed out that a full set of observations would take thirty years, he settled for a more reasonable agenda: an updated astronomical treatise (*zij*) and a new solar calendar and era. The poet and mathematician Omar Khayyam (1048–1132) was put in charge of the work. Although the *Zij-i Malik Shahi* was only a partially revised version of an earlier treatise, including some new observations and tables, the new solar calendar and era (*Tarikh-i Jalali*) was a major advance. Adopting new names for the months and days, the Jalali Era added an extra day every four years (leap year) and corrected the date of Nau Ruz – returning the first day of the year to the Vernal Equinox. In addition to casting horoscopes for Malik Shah, Omar was also responsible for important contributions to the field of mathematics – establishing algebra as a separate discipline from arithmetic and producing a commentary on Euclid.[43]

Observatory at Maragha

The observatory at Maragha, the new capital of the Ilkhanid ruler Hulaghu (r. 1256–65), was the most advanced in the world. Responsibility for its construction and organization rested with Hulaghu's chief *munajjim*, Nasir al-Din Tusi (1201–74). Like Omar Khayyam, Tusi was a polymath – philosopher, mathematician, astrologer, and astronomer. Unlike Malik Shah's observatory, however, Hulaghu's has not totally disappeared. Its instruments included a mural quadrant, an armillary sphere, a solsticial armilla, an equinoctial armilla, and a large domed building designed to measure the apparent diameter of the sun, moon, and five planets. Because their number, size, and sophistication required a sizable staff, Nasir al-Din, with the approval and support of Hulaghu, recruited a large group of astronomers – from the Islamic world and beyond (some as far away as China). He assembled a library – mostly astronomical, astrological, and mathematical – and founded a madrasa (college). Among the more distinguished scientists were Muin al-Din al-Maghribi (ca. 1220–82) and Nasir al-Din himself.

[43] H. Taqizadeh, "Various Eras and Calendars Used in the Countries of Islam," *BSOAS* 10 (1939), 110–12; Aydin Sayili, *The Observatory in Islam* (Ankara: Turk Tarih Kumumu Basimevi, 1988), 159–66; *Encyclopaedia of Islam*, 2d. ed., s.v. "al-Nudjum."

Muin al-Din, famed for his work in mathematical theory (especially trigonometry), left a record of the observations he made between 1262 and 1274. Nasir al-Din created trigonometry as a separate discipline, composed a detailed commentary on Ptolemy's *Almagest*, and wrote important works on medicine, minerals, philosophy, and ethics. In its day and for the next two hundred years, the *Zij-i Ilkhani* was the most complete and accurate astronomical treatise in the Eurasian world. The result of twelve years of observation, it was written first in Persian, then translated into Arabic. It contained corrections and additions to Ptolemy's *Almagest*, an updated star catalogue, and tables for computing the positions of the planets. Hulaghu's observatory at Maragha was also the first observatory to be funded by a permanent endowment (waqf) and, as a result, its work continued for another seventy years, well after Hulaghu's death. Tusi, however, like the other *munajjims*, had important astrological responsibilities as well. Hulaghu never made any important decision – personal or political – without consulting his chief astrologer.[44]

Observatory of the Time and Hour

Like Malik Shah's observatory, the institution founded by Rukn al-Din in the central Iranian city of Yazd has left no trace. According to a sixteenth-century history, it was completed in 1325 during the reign of the Ilkhanid ruler Abu Said Bahadur (1316–1325). In addition to the observatory, Rukn al-Din constructed a madrasa, a library (three thousand volumes), a mosque, and a hospital. Although the structure was called an observatory (the full name was Rasad-i Waqt-i Saat), it is probably more accurately understood as an exceptionally elaborate House of Chronology or Muwaqqit Khana, probably attached to the nearby mosque. The house contained a large water clock, like the one Harun al-Rashid sent to Charlemagne. It displayed the months of the year in various calendars (Greek, Turkish, Islamic, and Iranian), the hours of prayer, the days of the week, and the mansions of the moon. Like Hulaghu, Rukn al-Din dedicated a waqf for its upkeep, and, as a result, this institution, like the one at Maragha, outlasted its founder.[45]

[44] *Encyclopaedia of Islam*, 2d. ed., s.v. "'Ilm al-Haya"; *Encyclopaedia of Islam*, 2d. ed., s.v. "Zidj"; Sayili, *Observatory*, 187–223.

[45] Sayili, *Observatory*, 236–41.

Observatory at Samarqand

The final and most advanced observatory of the pre-1500 period was constructed near Samarqand by Ulugh Beg (1394–1449), the grandson of the great Central Asian conqueror Timur. Unlike Malik Shah or Hulaghu, Ulugh Beg was more interested in astronomy and mathematics than in astrology. In 1417 he broke ground, constructing a madrasa and a library and installing a sophisticated array of observational instruments – a large quadrant, a marble sextant, and an armillary sphere, among others. The first director of the madrasa was Jamshid Ghiyas al-Din al-Kashi, who recruited a group of seventy to one hundred astronomers and organized a series of formal lectures. In 1420 he completed the *Zij-i Khakhani*, a partial revision of Tusi's *Zij-i Ilkhani*. Another director of the observatory was Bursali Kadizade-i Rumi (d. 1440), an Ottoman *muneccim* who contributed a great deal to Ulugh Beg's new *zij*.[46]

Ulugh Beg, however, was a talented astronomer and mathematician, able (according to al-Kashi) to quickly calculate equivalent dates across the various calendars and eras. In 1437–1448 he completed the *Zij-i Sultani or Gurgani*, containing the first completely new star catalogue since Ptolemy's. He calculated the correct length of the solar year and obtained more accurate figures for the location and movement of the stars and planets. Translated into Latin, the *Zij-i Sultani* was an important influence on Tycho Brahe and Copernicus.[47]

The principal contribution of the Islamic observatory to the definition of time was the *zij* (astronomical treatise). Between the eighth and the nineteenth centuries, well over two hundred *zijs* were compiled in the Islamic world. The *Ain-i Akbari*, composed at the Mughal court in the late sixteenth century, listed eighty-six. The earliest were based on Indian and Iranian models, but from the ninth century onward the Ptolemaic tradition predominated. Only about twenty of the two hundred, however, were really new, based on fresh observations and calculations. These included the *Zij-i Malik Shahi*, the *Zij-i Ilkhani*, and the *Zij-i Sultani*.

Most *zijs* were intended to serve a single locality: a terrestrial longitude underlay the solar, lunar, and planetary tables and a terrestrial latitude

[46] E. Ihsanoglu, "Ottoman Science: The Last Episode in Islamic Scientific Tradition and the Begining of the European Scientific Tradition," in E. Ihsanoglu, *Science, Technology, and Learning in the Ottoman Empire* (London: Ashgate Publishing House, 2004), 21.

[47] *Encyclopaedia of Islam*, 2d. ed., s.v. "Zidj"; Sayili, *Observatory*, 260–89.

underlay the tables for spherical astronomy. The typical *zij* included: (1) an introduction to mathematical notation; (2) a chapter on chronology describing the relevant calendars and eras and the methods for converting dates from one temporal system to the other; (3) trigonometric tables; (4) spherical astronomical functions for solar or stellar timekeeping; (5) tables containing the motions, positions, equations and latitudes of the seven heavenly bodies; (6) planetary stations and visibility tables; (7) solar and lunar parallaxes and eclipses; (8) lunar visibility tables; (9) geographical tables containing the longitudes and latitudes of nearby towns and cities; (10) star catalogues; (11) tables for casting horoscopes; and (12) historical and ceremonial tables listing the dates of rulers and dynasties and the names and dates of various festivals, rituals, and ceremonies. A lengthy affair, the typical *zij* ran to one hundred pages or more.[48]

For the practicing *munajjim*, the *zij* was the basic reference work. Long, dense, and rarely updated, it was like an unabridged dictionary or an encyclopedia, not frequently consulted. Nevertheless, the *zij* was necessary – the annual almanac (Persian, *taqvim*; Turkish, *takvim*) could not be produced without its data, and it contained a mine of indispensable information on other subjects, for example, the geographical coordinates of nearby towns and cities, which were necessary for locating the *qibla* (direction of prayer) niche in newly founded mosques.

Although the early modern *munajjim* might occasionally consult the *Zij-i Sultani*, for his day-to-day work he would ordinarily turn to the annual *taqvim* (almanac). Produced by local *munajjims* at the beginning of the year, the *taqvim* extracted from the latest *zij* the relevant astronomical and astrological data. The booklet itself varied in size and splendor: Those for the rich and powerful were gilded, beautifully illustrated, and penned by a skilled calligrapher, whereas the ordinary householder had to be content with a smaller, shorter, undecorated version.

In the early modern period the ordinary *taqvim* had twenty-four pages, a two-page spread for each month of the year. It opened with a diagram of the seven heavenly bodies at the beginning of the solar year (the Vernal Equinox). The horoscope for the entire year followed, providing political,

48 *Encyclopaedia of Islam*, 2d. ed., s.v. "Zidj"; *Encyclopaedia of Islam*, 2d. ed., s.v."Tarikh"; North and Porter, *History of Astronomy and Cosmology*, 180–3; E. S. Kennedy, "A Survey of Islamic Astronomical Tables," *Transactions of the American Philosophical Society* 46 (1956): 123–77.

climatological, and agricultural forecasts. If major astronomical events were expected – eclipses, conjunctions, comets, or shooting stars – these were noted and interpreted.

On the right-hand side of the first page were columns listing the days of the week in the various calendars and eras (the Yazdegird, Jalali, Hijra, Julian, and often the Turkish Twelve-Year Animal). To the far right were annotations – specifying the relevant historical events (founding of a dynasty or birth of a prophet), festivals (of Muslims, Jews, Christians, and Hindus), or seasonal activities (beginning of the shipping season). On the left-hand side of the page was the ephemeris, a row of columns providing for each day the longitudinal position along the zodiacal ecliptic of the sun, moon, and five planets. Other columns contained the length of the day and the solar altitude at noon and at the afternoon prayer. At the far left the columns were given over to astrological matters. The relationships among the heavenly bodies were described and short predictions were given, identifying for that day which activities were favorable or not. Because the moon was so important in Islamic chronology, it was given a separate column, listing for each day its location and first visibility.[49]

Building on this brief introduction, the following chapters take up the transformation of the Islamic temporal system as it moved in time and space from the fledgling states of the seventh to thirteenth centuries to the early modern empires of the sixteenth to eighteenth. Before, however, plunging into an analysis of the Islamic concept of time – in its calendrical, ceremonial, and chronological dimensions – a short overview of the three empires is necessary. Although it is obviously out of the question to try and capture in a single chapter the detailed complexity of the secondary literature, without some sort of context a proper appreciation of the individual temporal transformations is impossible. From the calendrical point of view the principal question concerned the solar calendar or calendars added to the lunar Hijra. The different solar cultures – Zoroastrian in the Safavid Empire, Indic in the Mughal, and Christian in the Ottoman – gave birth to different roles for the astronomer/astrologer, different advances in the construction of observatories and the

[49] See Abu Rayhan al-Biruni, *Kitab al-Tahfim*, tr. R. R. Wright (London: Luzac, 1938), 186–91; Abu Rayhan al-Biruni, *Chronology of Ancient Nations*, trans. Edward Sachau (London: Oriental Translation Fund, 1879), 233–67; *Encyclopaedia of Islam*, 2d. ed., s.v. "takwim"; G. Saliba, "Computational Techniques in a Set of Late Medieval Astronomical Tables," *Journal of the History of Arabic Science* 1 (1977): 24–9.

compiling of astronomical treatises, and different achievements in timekeeping technology. From the ceremonial perspective, the issue was the annual round of rituals and celebrations. Whereas each empire inherited the four ceremonies of the early community, the expanded cycle that finally emerged reflected both the local variant of Islam as well as the indigenous demands for legitimacy and authority. Older rituals were dropped, traditional rites were revised and reworked, and entirely new ceremonies were introduced. From the chronological viewpoint the Islamic concept of time raised two topics – both unique to the early modern era. The first revolved around the problems created for an agrarian economy by a lunar era. In order to manage an agrarian fiscal system, early modern states were compelled to add one or more solar eras to the liturgical Hijra. Employing different chronological systems, however, was confusing, and the three states took different approaches to two problems: organizing the collection and disbursement of revenues and dating documents and histories. The other chronological issue was the appearance in the late sixteenth century of a cluster of millenarian prophets and movements, spurred on by the near-simultaneous occurrence of two momentous events: the Grand Conjunction of Jupiter and Saturn in 1583 and the end of the first Islamic millennium in 1591. The last chapter situates the Islamic concept of time in a wider context. Viewed from either the Judaic or the Christian perspective, the Islamic temporal system was a radical departure. Moreover, the revised early modern version (in the form described here) exerted an influence well beyond the confines of the three empires. The Islamic concept was an important influence on the temporal systems of early modern Europe.

1

Safavid, Mughal, and Ottoman Empires

The three Islamic empires of the early modern period – the Mughal, the Safavid, and the Ottoman – shared a common Turko-Mongolian heritage. In all three the ruling dynasty was Islamic, the economic system was agrarian, and the military forces were paid in grants of land revenue. Despite these similarities, however, significant differences remained. And, to fully appreciate the individual temporal systems, a brief description of the political, economic, religious, and cultural conditions in each state is necessary. Within the confines of a single chapter, however, it is not possible to review all of the literature and settle all of the controversies. As a result, the brief overview that follows depends, for the most part, on the most recent general histories and surveys.

SAFAVID EMPIRE (1501–1722)

Safavid Iran was shaped like a bowl, a flat bottom encircled by two mountain ranges. The Elburz Mountains ran along the southern shore of the Caspian Sea and met the smaller ranges of Khurasan in the east. The Zagros Mountains stretched from Azerbaijan in the northwest to the Persian Gulf and then east toward Baluchistan. The Eastern Highlands bordered the country on the southeast. A high arid plateau, with an average elevation of 3,000 feet, formed the base of the bowl. Two deserts – the Kavir and the Lut – sprawled across this expanse. Only three rivers interrupted the dry plateau: The Karun River (the only navigable one) originated in the Zagros Mountains and flowed to the Shatt al-Arab and the Persian Gulf; the Safid River rose in the Elburz Mountains and emptied into the Caspian Sea; and the Zayanda River, the only one of the three that

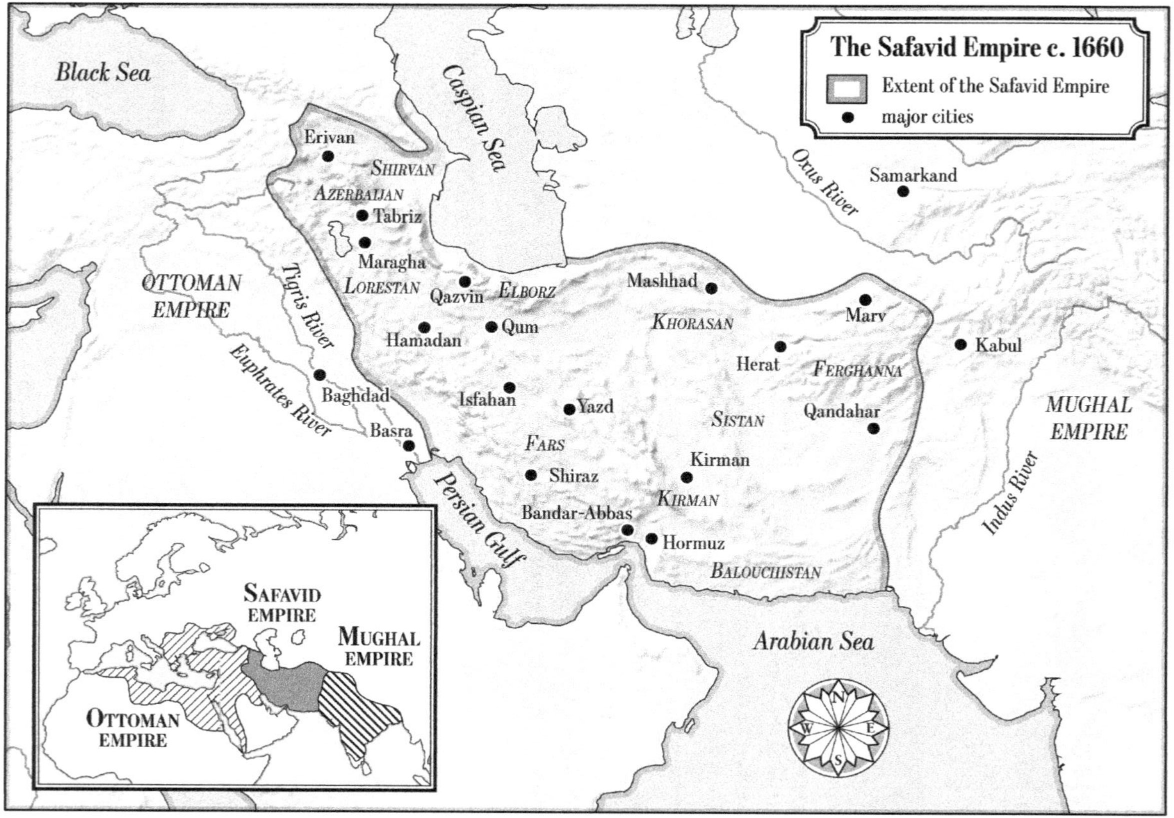

MAP 1. The Safavid Empire, c. 1660

watered the plateau, began in the Zagros Mountains and flowed through Isfahan, dying in a salty swamp nearby.

No reliable estimates are available for the population of Safavid Iran. However, given the scarcity of arable land, the total was considerably below that for either the Mughals or the Ottomans. Because of the large area taken up by mountains and deserts, only about one-eighth of the country was tillable.[1] As a result, an estimate for 1650 of eight to ten million seems reasonable.[2]

In 1501 Shah Ismail I (1501–1524), the founder of the dynasty, defeated the Turkish Aq Quyunlu forces that had, along with the Qara Quyunlu, ruled northwestern Iran since 1396.[3] Like his father and grandfather, Ismail headed the Safaviyya Sufi order. As Twelver or Imami Shiites, this mystical order rejected the first three caliphs and honored the Twelve Imams as the direct descendants of Muhammad. An invented genealogy claimed that Sheikh Safi (the founder of the order and Ismail's ancestor) was a lineal descendant of the Seventh Imam, Musa al-Kasim. Ismail also proclaimed himself the Mahdi (Guided One) and a reincarnation of Ali (the first Imam).

The Safavid founder united in his person the two ethnic components of the state – the Turkish Qizilbash "men of the sword" and the Persian Tajik "men of the pen." Except for their loyalty to Ismail and their membership in the Safaviyya order, the Qizilbash warriors were indistinguishable from their Sunni brethren in the eastern provinces of the Ottoman empire. Their common heritage gave the Safavid-Ottoman rivalry a special intensity. By contrast, as a descendant of landowners from the province of Gilan, Ismail also had a Persian side. Under him, as under the earlier Turkish rulers, Iranian scribes filled judicial, religious, and administrative positions. The hereditary notables, peasants, merchants, and artisans were Persian also. During Ismail's reign tension and competition marked the relationship between these two groups.

Shah Tahmasp I (1524–1576), eldest son of Ismail, ascended the throne at age ten. His first twelve years in power (1524–1536) witnessed a civil

[1] W. B. Fisher, ed., *Cambridge History of Iran, vol. 1: The Land of Iran* (Cambridge: Cambridge University Press, 1968), 611.

[2] Stephen P. Blake, *Half the World: The Social Architecture of Safavid Isfahan, 1590–1722* (Costa Mesa, CA: Mazda Publishers, 1999), 5.

[3] For a general overview, see Peter Jackson and Laurence Lockhart, eds., *The Cambridge History of Iran, vol. 6: The Timurid and Safavid Periods* (Cambridge: Cambridge University Press, 1986) and Andrew Newman, *Safavid Iran: Rebirth of a Persian Empire* (London: I. B. Tauris, 2008).

war between the Shamlu and Ustajlu Qizilbash tribes, and he had no real independence as a ruler. To impose order, Tahmasp introduced a program aimed at reestablishing imperial authority, a program extended and brought to conclusion by Abbas I. He created a new tribe, the Shahvand, and gave it a position equal to that of the other tribes. He increased the number of Qizilbash in his personal bodyguard and assembled a household troop of Christian slaves. He also divided each Qizilbash tribe internally and shifted tribal leaders from post to post. Some chieftains were kept at court in administrative positions while others were given military commands or provincial governorships. These measures weakened the powers of the tribes and laid the groundwork for the more thoroughgoing reforms of Abbas I.[4]

Like Ismail, Tahmasp was considered by his Qizilbash followers to be divinely favored. They gave him the messianic title Lord of the Age (Sahib-i Zaman). In 1557 he shifted his capital from Tabriz to Qazvin. The need to escape Ottoman threats in the Northwest and to more easily defend his borders in the Northeast probably prompted the move. Because Qazvin had long been a stronghold of Sunni orthodoxy, Tahmasp made a major effort to spread Shiism in his new capital. He decorated the mosques and madrasas with Shiite slogans, repaired the shrines of Shiite saints, commissioned religious poetry and funeral elegies, and expanded the ceremonies of religious mourning.[5]

On his accession at age sixteen, Shah Abbas (1587–1629) appointed his tutor, Murshid Quli Khan, viceroy.[6] To counter the threat of Qizilbash insurgency the young emperor executed a group of tribal chieftains, and, when he was strong enough, ordered the elimination of his tutor. Free at last, Abbas undertook a radical reorganization of the state. He greatly increased the number of cavalrymen in his personal bodyguard. These men, although they were Qizilbash tribesmen, differed from their kinsmen in their absolute loyalty to the shah: they left their tribal homelands, came to court, and became members of the imperial household. Under Abbas they expanded to between ten thousand and fifteen thousand men, and, by

[4] H. Roemer, "The Safavid Period," in Peter Jackson and Laurence Lockhart, eds., *The Cambridge History of Iran, vol. 6: The Timurid and Safavid Periods* (Cambridge: Cambridge University Press, 1986), 233–49; Masashi Haneda, *Le chah and les Qizilbas: Le Systeme militare safavide* (Berlin: Klaus SchwarzVerlag, 1987), 133–42; Richard Tapper, *The Frontier Nomads of Iran: A Political and Social History of the Shahsevan* (Cambridge: Cambridge University Press, 1997), 48–9.

[5] Newman, *Safavid Iran*, ch. 2.

[6] For a discussion see Newman, *Safavid Iran*, ch. 4.

the end of his reign, the highest-ranking held provincial governorships and state offices, and their leader had become the most important official in the state.[7] The emperor also created a corps of household slaves composed of Armenian, Georgian, and Circassian converts to Shiite Islam. Numbering ten thousand to fifteen thousand, these slave soldiers were even more dependent on Abbas than were the Qizilbash cavalry. Slaves rose to high ranks and by the end of Abbas's reign they, along with the leaders of the household cavalry, held most of the important imperial posts.

To pay for his military reorganization Abbas instituted a series of economic reforms. He redistributed agricultural land from the domain of the tribal chieftains to the domain of the imperial household, thereby giving him the money to pay his newly expanded bodyguard. At about the same time he began to make a greater use of the local Armenian merchants, primarily to market silk from the province of Gilan, which in 1592 had been incorporated into the imperial household. To further increase household revenues and to take advantage of the arrival of the European East India Companies (primarily the English and the Dutch), Abbas in 1619 established a monopoly over the sale and export of silk.[8]

The emperor's religious role, however, differed significantly from that of his predecessors. Because of the fourteen-year civil war (1576–1590) after the death of Tahmasp and the sheer passage of time, Abbas did not appear to receive the same veneration from his Qizilbash followers as had Ismail and Tahmasp. Also his personal beliefs were ambiguous – he flirted with the Nuqtavi heresy in his early years and showed an uncharacteristic interest in the teachings of the Christian missionaries. Like the Mughal emperor Akbar, however, who made several pilgrimages to the shrine of the Sufi saint Muin al-Din Chishti between 1562 and 1579 (one on foot), Abbas in 1601 burnished his reputation for piety by completing a forty-one-day pilgrimage on foot from Isfahan to the shrine of the Imam Riza in Mashhad. Like his predecessors, he promoted the spread of the Shiite mourning rituals, commemorating the martyrdom of the Imam Husain.[9]

7 *Encyclopaedia Iranica*, s.v. "Army. Safavids"; Haneda, *Le chah and les Qizilbas*, 146–86.

8 Edmund Herzig, "The Rise of the Julfa Merchants in the Late Sixteenth Century," in Charles Melville, ed., *Safavid Persia: The History and Politics of an Islamic Society* (London: I.B. Tauris, 1996), 305–22; Willem Floor, "The Dutch and the Persian Silk Trade," in ibid, 323–68; Rudolph Matthee, *The Politics of Trade in Safavid Iran: Silk for Silver, 1600–1730* (Cambridge: Cambridge University Press, 1999).

9 Jean Calmard, "Shi'i Rituals and Power II: The Consolidation of Safavid Shi'ism: Folklore and Popular Religion," in Charles Melville, ed., *Safavid Persia: The History and Politics of an Islamic Society* (London: I.B. Tauris, 1996), 139–90; Charles Melville, "Shah Abbas and the Pilgrimage to Mashhad," in Ibid., pp. 191–229.

At the beginning of his reign, Abbas made peace with the Ottomans on disadvantageous terms (ceding the province of Azerbaijan and Tabriz, its capital) in order to concentrate his forces against the Uzbeks in the Northeast. In 1590 he transferred his capital from Qazvin to Isfahan, and in 1598, his military reorganization underway, Abbas reconquered Herat and Mashhad. Having achieved peace in the east, the emperor turned his attention to the Ottomans and in 1603–1604 recaptured Azerbaijan. The treaty of 1612 reestablished the old boundaries between the two states and in 1622, with help from the English, he expelled the Portuguese from Hormuz.

Abbas died in Mazandaran in 1629 and was succeeded by his eighteen-year-old grandson, Shah Safi (1629–1642). The coronation ceremonies included rituals from the Safaviyya order, suggesting that even at that late date Ismail's role as Sufi master had not been forgotten. Under Shah Safi, Abbas's policy of strengthening the imperial household at the expense of the tribal chieftains continued. Generous to his supporters but suspicious of potential rivals, the new emperor executed many high-ranking officials during the early years of his reign. Imam Quli Khan, Abbas's rich and powerful governor of Fars, was put to death in 1632, and his province added to the imperial domain. Raised in the harem, Safi had little interest in ruling. In 1633 he turned over management of the empire to his grand wazir, Mirza Muhammad Taqi, known as Saru Taqi. A man of honesty, integrity, and ability, he held the post until 1645, when he was assassinated by jealous rivals.[10]

Shah Safi also followed Abbas's lead in economic reorganization. He added the lands of the defeated Qizilbash chieftains to the domain of the imperial household. He rescinded Abbas's silk monopoly and, as a result, trade with the European companies increased dramatically. He also encouraged the spread of Shiism, witnessing the massive processions from the upper gateway of the imperial palace.[11]

During Safi's reign military conflict with Iran's neighbors recommenced. In 1629 the Ottomans captured Hamadan, but in 1630 the Safavids resisted their attempt to regain Baghdad. Sultan Murad IV captured Erivan and overran Tabriz in 1635. In 1639 the Ottomans recaptured Baghdad for the last time, and a treaty in the next year established

[10] Willem Floor, "The Rise and Fall of Mirza Taqi, The Eunuch Grand Vizier (1043–55/1633–45): Makhdum al-Omara va Khadem Al-Foqara," *Studia Iranica* 26 (1997): 237–66; Jean Chardin, *Voyages du Chevalier Chardin, en Perse* ..., 10 vols. (Paris: Le Normant, 1811), 7, 303–17; Newman, *Safavid Iran*, ch. 5.

[11] Calmard, "Shi'i Rituals and Power II," 139–90.

peaceful boundaries between the two states. By contrast, the skirmishes in the Northeast with the Uzbeks continued throughout the seventeenth century. Except for a brief battle over Qandahar, the relationship with the Mughals remained peaceful. When Safi died in 1642, the country was at peace. He was buried in Qum.

Abbas II (1642–1666) succeeded his father to the throne at age ten.[12] At this point the Qizilbash chieftains had lost their preeminent position in the state, having to contend with the leaders of the imperial household troops and the ranking clerics. After Saru Taqi's murder, Abbas II began to play a more active role in state affairs and devoted several days a week to administration. Internationally, the truce with the Ottomans continued, and no full-scale battles with the Uzbeks erupted.[13]

Although he had no significant role in the Safaviyya Sufi order, Abbas II, like the other emperors after Ismail, retained a reputation for sanctity. He continued to promote the spread of Shiite ceremonies and festivals but the popular interest in messianic, esoteric sects persisted. The 1639 treaty with the Ottomans put an end to the skirmishes on the western borders, and the 1657 trade agreement spurred an upsurge in commercial activity. Economic problems, however, could not be eliminated: Abbas declared a tax amnesty and underwrote several large diplomatic receptions, he fought the Mughals over Qandahar, and he carried out an extensive building program in Isfahan. To deal with the revenue shortfall Muhammad Beg, the new wazir, raised taxes, devalued the currency, and reduced the size of the army. Although not uniformly successful, these reforms seemed to restore a measure of economic stability.[14]

On the death of his father, the twenty-year-old Safi Mirza (1666/68–1694) came to the throne. Crowned in 1666 as Safi II, the new emperor faced so many serious problems (bad harvests, an earthquake, and Cossack raids) that his ministers decided that his coronation had been ill-fated. The imperial astrologers chose a new date, and in the second ceremony two years later he took the name Suleiman. His reign was relatively peaceful – no battles with the Ottomans, Uzbeks, or Mughals.[15]

The new enthronement, however, didn't completely change Suleiman's luck. Natural disasters – harsh winters, swarms of locusts, drought, and

12 Newman, *Safavid Iran*, ch. 6.

13 Roemer, "Safavid Period," 288–304.

14 Willem Floor, "The Dutch and the Persian Silk Trade," 323–68; Rudi Matthee, "The Career of Mohammad Beg, Grand Vizier of Shah `Abbas II (r. 1642–66)," *Iranian Studies* 24 (1991): 17–36.

15 Newman, *Safavid Iran*, ch. 7.

bad harvests – plagued his entire reign. The promotion and elaboration of Imami Shiism continued: important works on the theology and philosophy of the creed appeared while the court underwrote an expansion of Ashura ceremonies and the devotions at the shrines of local saints.[16]

Because Suleiman had not arranged for a successor, a series of harem intrigues put Sultan Husain (1694–1726) rather than his younger brother, Abbas Mirza, on the throne.[17] By this point, however, the empire was at peace, not threatened by any of its traditional enemies. The imperial administration, refined and developed over the years, was able to handle the routine problems of governance. Economic problems persisted: crop failures brought on inflation and occasional famine, and the trade deficits with India and the European companies led to attempts to control the outflow of specie. Sultan Husain's building projects – the Chahar Bagh complex and the Farahabad garden palace – put a considerable strain on the imperial treasury.

Quiet, studious, and absorbed in religious matters, the Shah fell under the influence of several influential clerics, most notably the religious scholar Mir Muhammad Baqir Khatunabadi, the first Mulla Bashi. In 1695 the emperor promulgated an edict signed by the Sheikh al-Islam of Isfahan (Majlisi II) and six other prominent ulama banning all non-Islamic activities. In the central square of the capital, six thousand bottles of Georgian and Shirazi wine from the imperial cellars were smashed.[18]

Muhammad Beg's policy of neglecting the military, having been followed by Suleiman and then Sultan Husain, finally had its effect. Although the army was able to meet the usual challenges in the first decade of the eighteenth century, the Afghan attacks in the second and third decades seemed to catch the Safavid generals by surprise. In 1711 the Ghilzai Afghans captured Qandahar and in 1721 they arrived outside Kirman. In 1722, after a long and terrible siege, the Afghans crushed the larger Safavid army and sacked Isfahan. Shah Sultan Husain was beheaded in 1726, and the two figurehead shahs who prolonged the dynasty were replaced in 1736 by Nadir Shah, a Turkman of the Afshar tribe.

[16] Roemer, "Safavid Period," 304–10; Rudi Matthee, "Administrative Stability and Change in Late-17th-Century Iran: The Case of Shaykh Ali Khan Zanganah (1669–89)," *IJMES* 26 (1994): 77–98; Calmard, "Shi'i Rituals and Power II," 164.

[17] Newman, *Safavid Iran*, ch. 8.

[18] Kathryn Babayan, "Sufis, Dervishes and Mullas: The Controversy over Spiritual and Temporal Dominion in Seventeenth-Century Iran," in Charles Melville, ed., *Safavid Persia: The History and Politics of an Islamic Society* (London: I. B. Tauris, 1996), 117–38.

MUGHAL EMPIRE (1526–1739)

In both area and population the Mughal empire was by far the largest of the three states.[19] In 1650 it boasted a population of about 150 million people and covered nearly the entire Indian subcontinent. From the beginning of the first millennium CE, India had been an Eldorado, famed throughout the Eurasian world for its spices, textiles, diamonds, and paper. As an agrarian empire its size and wealth were heavily dependent on its climate. Unlike the arid, sparsely populated Safavid and Ottoman empires, Mughal India was tropical. The two great river systems (the Indus and Ganges) and the annual monsoon sustained a remarkably rich growing season, yielding two bumper crops a year (primarily wheat and rice). In the north the subcontinent was sheltered from the polar winds of Central Asia by the Himalaya and Hindu Kush mountains. From these ranges flowed the two rivers that watered the north Indian plains. The Jamuna-Gangetic system flowed south and east emptying into the Bay of Bengal while the Indus system (comprising the five rivers of the Punjab) ran west and south into the Arabian Sea. The southern peninsula of the subcontinent was cut off from the North by the Vindhya Mountains. Although at ten thousand feet they were not as forbidding as the towering northern ranges, they constituted a significant natural barrier, dividing the peoples and cultures of India into two very different halves.

In addition to the river systems, the other major contributor to India's size and wealth was the monsoon, a season of torrential rains that inundated the subcontinent from two directions. The first blew in from the Arabian Sea in early June, watering the southwest coast and moving eastward across most of the country by the first week of July. The second, originating in the Bay of Bengal, spread over Assam and was deflected by the Himalayas into North India, arriving in early July also.

Under the Mughals this rich, culturally complex, and heavily populated region was slowly molded into a functioning state. Like other north Indian empire builders, the early Mughals established their headquarters near the conjunction of the Ganges and the Jamuna. Having consolidated themselves in the north, they extended their control to the eastern and western edges of the flood plain and then moved beyond the Vindhyas, gradually incorporating the lands of central and southern India into their evolving empire.

[19] For a general introduction, see John Richards, *The Mughal Empire* (Cambridge: Cambridge University Press, 1993).

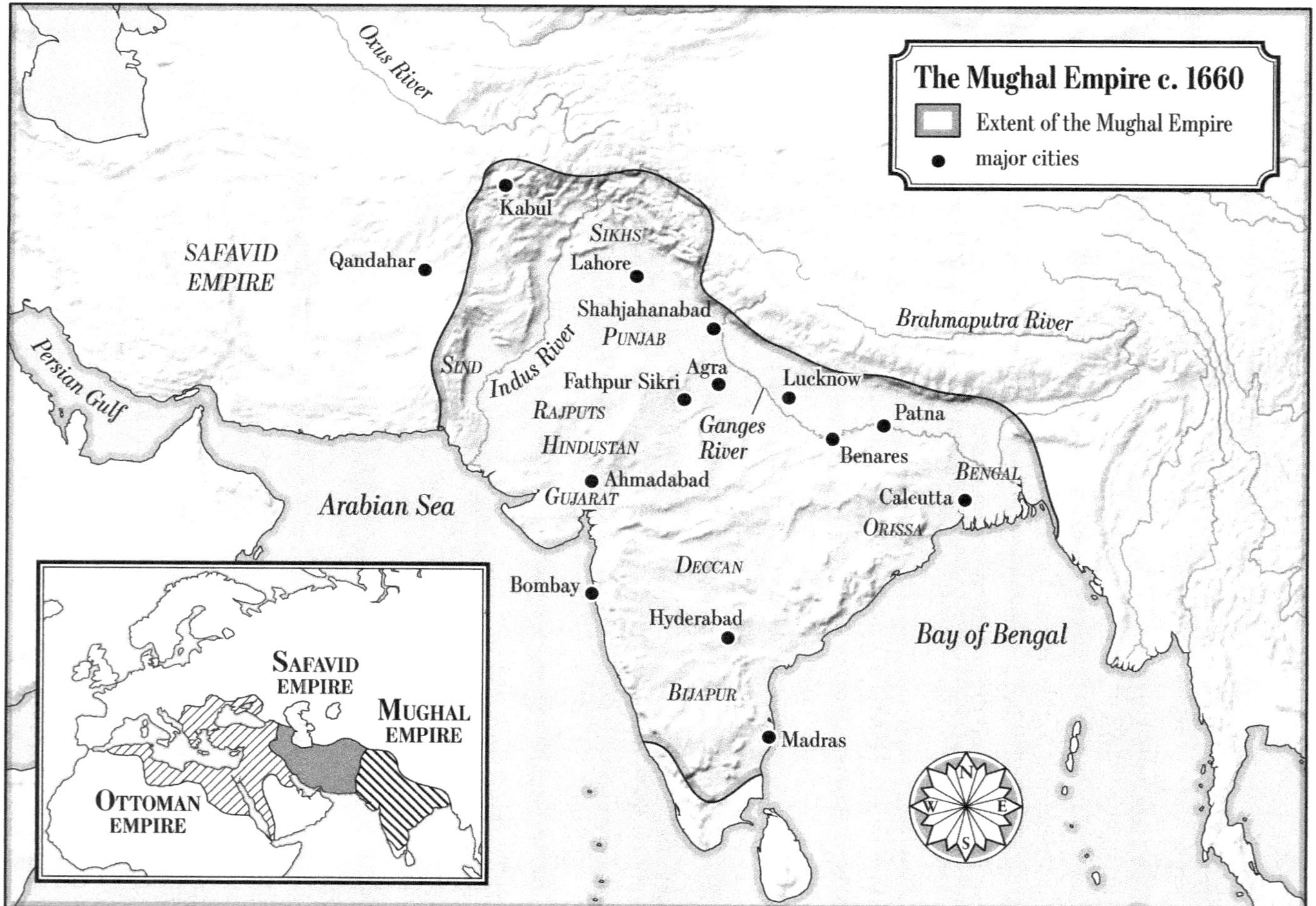

MAP 2. The Mughal Empire, c. 1660

Muhammad Zahir al-Din Babur, a Chagatai Turk from Fergana in Central Asia, was the founder of the Mughal Empire. Although Babur could trace a connection to Chagatai Khan, the second son of Chinghiz Khan (ca. 1162–1227), through his mother, it is by no means accurate to call him or his successors Mongol. Mughal, the name of the dynasty, is a variant of Mongol and was used in India to distinguish immigrants or the recently immigrated from local Muslims. Because Babur's father, Umar Sheikh Mirza, was directly descended from Timur (1336–1405), the great Central Asia empire builder, it is more accurate to call the dynasty Timurid, the name by which it was known to Indians of the period.

In 1526 Babur, at that time ruler of a city state centered on Kabul, defeated the Afghan rulers of North India and inaugurated Mughal rule in the subcontinent. He immediately captured Delhi (later Shahjahanabad), Agra, Gwalior, and Kanauj and in 1527 defeated the massed armies of the Rajput ruler, Rana Sangha. By 1529 he was master of the Indo-Gangetic Plains all the way to Patna but in 1530, at the height of his power, he died.

Humayun (1530–1556), Babur's son and successor, faced a difficult task. He had to mold territories in Afghanistan, Punjab, and the Gangetic Plains into a functioning state, and he had to do it against the opposition both of his own followers and of the recently defeated Afghans. It is no wonder that he failed and was forced to seek refuge with the Safavid ruler of Iran, Shah Tahmasp. From 1540 until Humayun's return to India in 1556 Afghans ruled North India.

Jalal al-Din Akbar (1556–1605), like the Safavid ruler Shah Abbas I, came to the throne as a callow, untested teenager (Akbar at thirteen, Abbas at sixteen) and, like Abbas, had to rid himself of an overbearing tutor and successive challenges to his authority. In 1571 having asserted his authority over his fractious followers and defeated his principal north Indian rivals, Akbar moved his headquarters from the old north Indian capital of Agra to a new imperial center named Fathpur Sikri, some twenty-four miles to the west. From there he launched a far-reaching campaign of radical reform.

Akbar's new imperial order was the result of three kinds of reform – military-administrative, economic, and cultural. The first problem facing Akbar was how to organize, pay, and ensure the loyalty of his martial followers? The young ruler had witnessed firsthand his father Humayun's difficulty maintaining a group of reliable commanders, men who could be counted on in both peace and in war. In 1575 he instituted a program of branding, requiring the horses of each cavalryman to carry two brands: his captain's and the emperor's. Soon after, Akbar began to make the first appointments in what was to develop into the characteristic feature of Mughal rule – the *mansabdari* or officeholder system. This was a ranked military-administrative hierarchy, each member of which filled an administrative or military position and provided a certain number of armed and mounted followers. At about the same time, Akbar began to reorganize his administration. He established a daily routine for dealing with military and economic matters,[20] set up a record office,[21] divided the empire into provinces,[22] and ordered a village by village census (probably never completed).[23]

For the early modern Islamic world Akbar's new military-administrative system was unusually open to ethnic and religious differences. Unlike the Ottoman and Safavid states, where conversion was required, the Mughals decided not to restrict membership in the *mansabdari* system to the Central Asian Sunni warriors who had made up the bulk of Babur's followers and who had accompanied Humayun on his reconquest of India in 1555. Rather, Akbar's system included all of the local martial groups: Turanis (Turkish-speaking Sunni Muslims from Central Asia), Iranis (Persian-speaking Shiites from Safavid Iran), Afghans (Sunni Muslims from eastern India), Sheikhzadas (Indian-born Muslims), Rajputs (Hindu warriors from north India), and Marathas (Hindu warriors from western India).

In addition to military-administrative restructuring, Akbar devoted a good deal of time to economic reorganization. Because the Mughal empire, like the Ottoman and Safavid states, was agrarian-based (with, to be sure, dynamic commercial, manufacturing, and financial sectors), this involved, for the most part, a reordering of the land revenue system. Assessments were out-of-date and numerous disagreements had arisen between imperial recordkeepers and army officers. Thus, in early 1575 Akbar ordered a detailed survey of central north India – measuring the arable land (establishing standard measurements for length and area) and collecting information on prices and yields. With this information the imperial revenue administrators in 1580 published the Ten-Year Settlement (Ain-i Dahsala),

[20] Abu al-Fazl Allami, *The A`in-i Akbari*, ed. D.C. Phillott, trans. H. Blochmann, 3 vols. (Calcutta: Bibliotheca Indica, 1927–1949; reprint ed., Delhi: Low Price Publications, 1989), 1: 162–5; Abu al-Fazl Allami, *Akbar Nama*, trans. H. Beveridge, 3 vols. (Calcutta: Bibliotheca Indica, 1948), 3: 256–7.

[21] A.L. Srivastava, *Akbar the Great*, 2 vols. (Delhi: Shiva Lal Agarwal and Company, 1962–68), 1: 178; *Akbar Nama*, 3: 118.

[22] Srivastava, *Akbar*, 2: 113; *Akbar Nama*, 3: 282.

[23] Srivastava, *Akbar*, 1: 284–5; *Akbar Nama*, 3: 95, 165–81, 346–7.

establishing a revenue rate in cash for each piece of land in the central empire. Thereafter, the members of Akbar's *mansabdari* system, paid in land revenue grants (or *jagirs*), were called *jagirdars*.

At about the same time Akbar also began to fashion a multifaceted imperial ideology, one that would foster a deeper commitment to him and his dynasty and that would also be hospitable to the religious beliefs of all his subjects. In 1575 he erected the Ibadat Khana (House of Religious Assembly) in Fathpur Sikri.[24] At first the discussions in this hall were traditional – the representatives were Muslims and the topics were Islamic beliefs and practices.[25] These sessions (from ca. 1575–1579), however, proved ultimately disillusioning. As he had been exposed during the previous twenty years to the religious diversity of early modern India, Akbar could not defer to the unimaginative religious specialists of traditional Islam. Deeply dissatisfied with the insularity of these men and their wholesale condemnation of nontraditional and non-Islamic beliefs and practices, Akbar decided to promote a new policy. Called "*sulh-i kull*," this new approach was developed during the period 1579–1582 with the help of Abu al-Fazl, Akbar's chief historian and ideologue, and his father, Sheikh Mubarak.[26] Usually translated "universal peace" or "absolute toleration," the phrase, it seems to me, is better rendered "lasting reconciliation." Akbar's intent was not to establish perfect harmony among the competing religious and cultural groups of the Indian subcontinent but, rather, to achieve a kind of *modus vivendi*.

"*Sulh-i kull*" was aimed at two quite different audiences: the one non-Muslim and the other Muslim. During the 1579–1582 period, Akbar became very interested in the non-Muslim religious traditions of the subcontinent – Hinduism, Jainism, Christianity, and Zoroastrianism. Because India was an overwhelmingly Hindu country and because Akbar had already decided to draw the Rajputs into the *mansabdari* system, he had begun as early as 1562 to marry the daughters and nieces of the Rajput chieftains. After 1579 Hindu mystics and Brahmin priests began to

[24] For a discussion see Stephen P. Blake, "Religious Conflict in Early Modern India: Akbar and the House of Religious Assembly," in Marguerite Ragnow and William D. Phillips Jr., eds., *Religious Conflict and Accommodation in the Early Modern World* (Minneapolis, MN: Center for Early Modern History, 2011), 69–82.

[25] Srivasta, *Akbar*, 2: 204–5; *Akbar Nama*, 3: 158–9.

[26] The author of *Mujmal Mufassal*, written during Shahjahan's reign, states that in 1579–1580 Akbar adopted the principle of "*sulh-i kull*"; S. A. A. Rizvi, The *Religious and Intellectual History of the Muslims in Akbar's Reign* (Delhi: Munshiram Manoharlal, 1975), 409.

frequent the discussions in the Ibadat Khana, and he began to appoint high-ranking Rajputs to important state offices. Although Akbar had always allowed his Rajput wives to follow their own customs, in the 1580s in Fathpur Sikri he began to participate in their religious ceremonies and rituals, commemorating the Hindu festivals of Diwali, Dussehra, Vasant, and Holi. Of the other three non-Muslim religious traditions, Zoroastrianism had the greatest impact on the emperor.

The second part of Akbar's "lasting reconciliation" policy was directed toward Muslims. In June 1579 he read the *khutba* (Friday sermon) in the central mosque of Fathpur Sikri. Although the accounts of his performance differ, this reading marks the beginning of Akbar's efforts to bring order to the contentious religious environment of *Muslim* India. The second step followed closely: Two months later in August Abu al-Fazl and his father drafted a *mahzar* (a document attested to by others) that proclaimed Akbar the adjudicator of religious disputes – either those between Sunnis and Shiites or those among the representatives of the four Sunni law schools.

The third, last, and most controversial part of Akbar's effort to bring order to Indian Islam was the Sufi-like imperial order that he founded – the Tauhid-i Ilahi (Divine Monotheism).[27] Most of the early, high-ranking members of the new order were Muslim (eighteen of the nineteen named in the *Ain*), and its organization and ceremony were modeled after the Sufi mystical orders of north India. In the *Ain-i Akbari*, volume three of the *Akbar Nama*, Abul al-Fazl's monumental history of Akbar's reign, Akbar's new order is discussed under the rubric "rules for the disciples [*ain-i iradat-i guzinan*]."[28] "*Iradat*" was the Sufi term for discipleship, and in the contemporary sources the members of the order were referred to as "disciples" ("*murids*") and their relationship with Akbar as "discipleship" ("*muridi*").[29]

Akbar expected the members of the Tauhid-i Ilahi to be in the vanguard of the "lasting reconciliation" movement, reflecting in their words and deeds a tolerance of their fellow Muslims and an appreciation for the

[27] *Ain*, 1: 211.

[28] Abu al-Fazl, *The A`in-i Akbari*, ed. H. Blochman, 2 vols. (Calcutta: Asiatic Society of Bengal, 1872–1877), 1: 7; *Ain*, trans. Blochman, 1: 175.

[29] In 1588, for example, a high-ranking *mansabdar* stamped his *farman* (order) with a seal reading "... the disciple [*murid*] of Akbar Shah ..." Rizvi, *Religious and Intellectual*, 406. Abu al-Fazl described one of the nobles as donning the "chain of discipleship [*muridi*]." Ibid., 399. Azam Khan and Mirza Aziz Koka joined the order and became *murids*. Ibid., 427–8.

cultural complexity of Mughal India. For example, the emperor pointed with approval to a recently immigrated Iranian who acted as if there were no difference between Sunnis and Shiites, following the principle of "lasting reconciliation or *sulh-i kull*."[30] Later, he admonished his son Prince Daniyal "be not offended by diversity of religion. Struggle hard to sit in the shade of "*sulh-i kull*."[31] Finally, in 1594 he sent a letter to Shah Abbas counseling tolerance and restraint. The Safavid ruler was at the end of a bloody campaign against the millenarian, extremist sect of the Nuqtavis. Akbar, however, had welcomed a leading member of the group to his court in 1577 and had sent a letter of support to another leader in 1584. Akbar wrote: "He [Shah Abbas] must . . . exercise supreme caution before putting any one to death and destroying what is an edifice of God. . . . It must be considered that the Divine mercy attaches itself to every form of creed, and supreme exertions must be made to bring oneself into the ever vernal flower garden of "*sulh-i kull*."[32]

Akbar's son and successor, Jahangir, wrote of his father's tolerance:

> The Professors of various faiths had room in the broad expanse of his incomparable sway. This was different from the practice in other realms, for in Iran there is room for Shias only, and in Turkey, India, and Turan there is room for Sunnis only . . . in his dominions . . . there was room for the professors of opposite religions, and for beliefs good and bad, and the road to altercation was closed. Sunnis and Shias met in one mosque, and Franks and Jews in one church, and observed their own forms of worship. "*Sulh-i kull*" was his disposition. He associated with the good of every race and creed and was gracious to all in accordance with their condition and understanding.[33]

Jahangir (1605–1628), the least forceful of the four great emperors, has usually been seen as weak and uncertain, failing to build on Akbar's successes and ceding much of his authority to his wife Nur Jahan. During his reign no serious attempt was made to extend Mughal dominion in the Deccan and South India, and Qandahar in central Afghanistan was lost to the Safavids. By contrast, Mughal rule in the province of Bengal was reorganized and put on a peaceful, stable footing. The number of

30 *Akbar Nama*, 2: 35.

31 Ibid., 3: 1079.

32 Ibid., 3: 1012.

33 Jahangir, *The Tuzuk-i Jahangiri*, ed. Henry Beveridge and trans. Alexander Rogers, 2 vols. (Delhi: Munshiram Manorharlal, 1968) 1: 37–8. In the translation, the sentence on *sulh-i kull* was omitted. See Ahsan Raza Khan, "Abu al-Fazl's Account of Akbar's Expansionism: Ambit of Reason and Tolerance?" (paper presented at the International Seminar "Reason and Tolerance in Indian History," New Delhi, India, October 2006), 13.

mansabdars expanded from about eight hundred to nearly three thousand, proving to be a major burden on the treasury and causing the percentage of state revenues controlled by the imperial household to drop precipitously.

The Emperor Shahjahan (1628–1658) was a different man altogether. Energetic, bold, and a skilled general, he readopted Akbar's policy of vigorous expansion. His first move was to reestablish Mughal rule in the Deccan. He was also responsible for the last serious attempt by the Mughals to recover Qandahar – winning it briefly, losing it to the Persians, and then failing on three separate occasions to regain it. By the middle of his reign, he had consolidated Mughal rule in most of the subcontinent. All this meant that he was free to patronize the arts – poetry, painting, and especially architecture. Shahjahan is best known as the builder of the Taj Mahal, that beautiful memorial to his wife in Agra, but he also renovated the palace-fortresses in Agra and Lahore and planned and built a new capital city (Shahjahanabad) in the Delhi area.

Aurangzeb (1658–1707), the last of the four great emperors, is an enigma. Possessed of energy, talent, experience, and discipline, he should have been the perfect ruler, presiding over a reign of peace and prosperity. Yet there is almost universal agreement that Aurangzeb was a failure and that his reign marked the beginning of the end. Like other Mughal princes before him, Aurangzeb grew discontented and revolted against his father. Unlike the others, however, he was successful, and in 1658 he locked his father in the Agra fort and replaced himself on the throne. A skillful general and a careful administrator, the new emperor brought Assam and Eastern India into the empire, subdued the Sikhs (a militant religious movement centered in the Punjab), and moved against the Marathas. His Maratha campaign, initially successful, soon bogged down, and he left North India in 1679 to direct the military effort in person. For twenty-eight years, until his death in 1707, he pursued the wily Maratha horsemen from place to place, conquering and reconquering small forts, fighting innumerable skirmishes, but always failing to force the one major battle that would have decided the issue. By temperament Aurangzeb was traditional, conservative, and compulsive, trying to roll back Akbar's policy of "lasting reconciliation" and unwilling to delegate administrative and military details.

Aurangzeb's son, Bahadur Shah (1707–1712), was an old man when he finally came to the throne, and the empire he inherited had been bled of men and resources by the long unsuccessful Deccan campaign. The reign of Farrukhsiyar (1713–1719), Bahadur Shah's successor, was undistinguished, and he was eventually replaced by Muhammad Shah (1719–1748). Like his immediate predecessors, Muhammad Shah had no interest in generalship or administration, devoting himself to hunting and palace amusements. In 1739 Nadir Shah, the newly crowned ruler of Iran, took Qandahar and Kabul from the Afghans and entered the subcontinent. Easily defeating the disorganized and badly led Mughal troops, he occupied Shahjahanabad. After a group of young toughs attacked and killed some nine hundred of his soldiers, Nadir ordered a general massacre. When the Iranian ruler finally left, the city lay devastated and the Mughal empire was, in any meaningful sense, at an end.

OTTOMAN EMPIRE

The Ottoman empire (ca. 1300–1923) was the first and longest-lived of the three early modern Islamic empires. Unlike the other two, it had no natural boundaries and controlled no coherent geographical entity. It was simply the "the domains of the House of Osman." Nevertheless, by the late sixteenth century the Ottomans controlled an enormous swath of territory: Anatolia, Iraq, the Balkans, Hungary, Syria, Egypt, the Arabian Peninsula, and North Africa. Unlike the heavily watered plains of north India or the arid highlands of Iran, the Ottoman empire encompassed a wide variety of climates – the lush Tigris-Euphrates and Nile deltas, the deserts of Arabia, and the more temperate climates of Anatolia, Syria, and North Africa. In 1600 the population of the empire was about twenty million.[34]

Osman (1281–1326), the eponymous founder of the dynasty, began as the ruler of a small Seljuq successor state in western Anatolia.[35] He and his early followers were Muslim Turks, descendants of the Central Asian Turkish tribes who migrated south, defeated the Abbasids at Baghdad

[34] Charles Issawi, "The Ottoman Empire in the European Economy," *Journal of the Economic and Social History of the Orient* 1 (1957): 109–17.

[35] For the most recent overviews, see Suraiya N. Faroqhi, ed., *The Cambridge History of Turkey, vol. 3: The Later Ottoman Empire, 1603–1839* (Cambridge: Cambridge University Press, 2006); Suraiya N. Faroqhi and Kate Fleet, eds., *The Cambridge History of Turkey, vol. 2: The Ottoman Empire as a World Power, 1453–1603* (Cambridge: Cambridge University Press, 2012; and Halal Inalcik and Donald Quataert, eds., *An Economic and Social History of the Ottoman Empire, 1500–1914* (Cambridge: Cambridge University Press, 1994). The older surveys by Stanford Shaw and Halil Inalcik are still valuable. Stanford Shaw, *History of Ottoman Empire*, 2 vols. (Cambridge: Cambridge University Press, 1967–77); Halil Inalcik, *The Ottoman Empire: Classical Age: 1300–1600* (London: Weidenfeld, 1973).

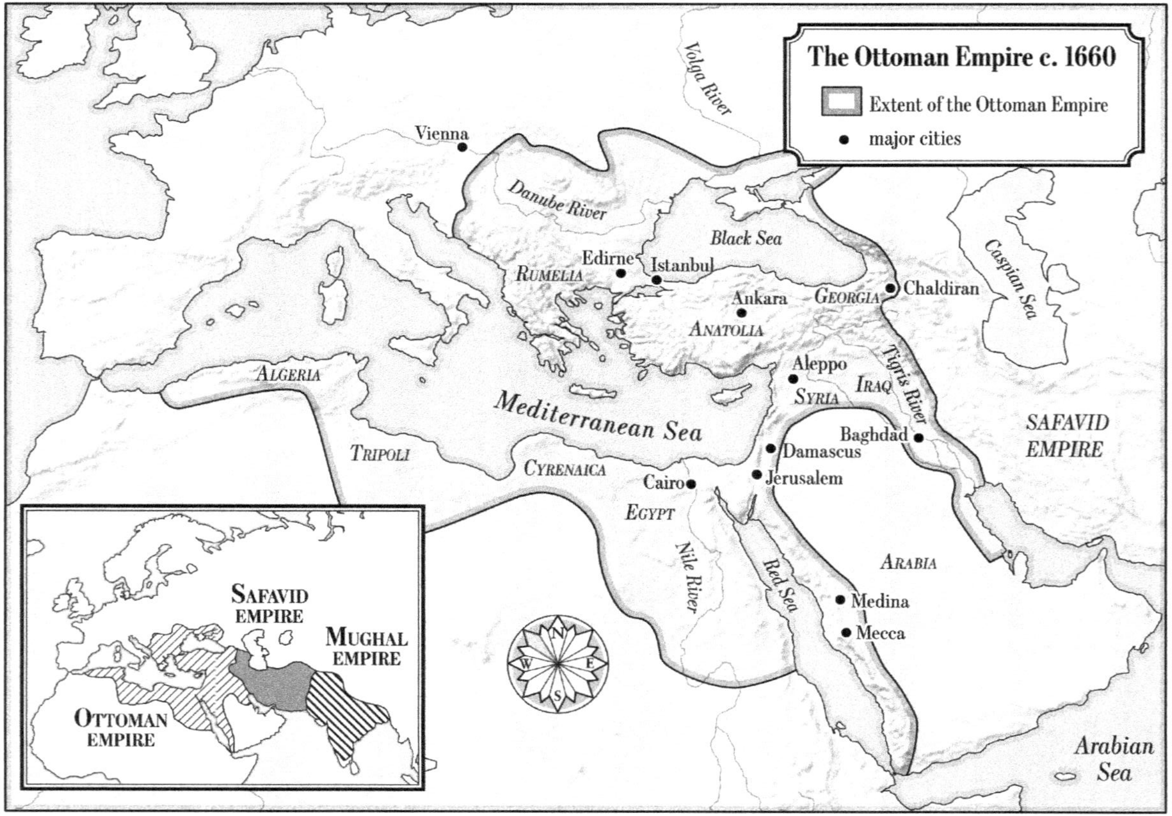

MAP 3. The Ottoman Empire, c. 1660

(1055), and established the Seljuq dynasty that ruled large parts of Anatolia, Iraq, and Iran from the mid-eleventh to the mid-fourteenth centuries. The descendants of these Central Asian tribesmen comprised the majority of Osman's followers in the late thirteenth and early fourteenth centuries and the bulk of Shah Ismail's warriors in the early sixteenth century. In the fourteenth century Osman and his successors slowly expanded their fledgling state – east across Anatolia and west into the Balkans. Under Bayezid I (1389–1402), however, the push eastward brought the Ottoman warriors up against the powerful forces of Timur. At the battle of Ankara in 1402 Bayezid's army was crushed. He was taken captive and died soon after. A major cause of the Ottoman defeat, foreshadowing a problem with the Safavids a century later, was the divided loyalties of Bayezid's men. His tribesmen, recognizing many of their old comrades among Timur's forces, soon defected to the enemy. A hundred years later the descendants of these men were again torn – this time between their loyalty to their Ottoman commanders, on the one hand, and the charismatic leadership of Shah Ismail, on the other.[36]

In the Ottoman army and imperial household slaves or servitors (*kul*) filled many positions. Although earlier Islamic regimes (the Abbasids, Seljuqs, and Mamluks, for example) had employed slaves, the Ottomans relied on them to a much greater extent that did the rulers of either the Mughal or Safavid empires. Because, according to Islamic law, Muslims could not be enslaved, most slaves were prisoners of war, employed primarily as soldiers. However, as the demand for slaves increased in the late fifteenth century, the Ottomans instituted a levy (*devsirme*). Considered an extraordinary tax on the non-Muslim cultivating families of the realm, it was ordered every three to seven years. In the sixteenth century the annual totals ranged from about one to three thousand boys. The young men were circumcised, converted to Islam, and taught Ottoman Turkish, but they were not mistreated – physically abused, restricted to menial occupations, or passed from owner to owner. Most of them came from the Balkans and were destined for the Janissaries, the sultan's personal infantry. A small percentage of the talented were sent to the palace school – entering imperial service and becoming eligible for the highest military and administrative offices. Among the Ottomans the slave system permeated the entire society, from top to bottom, and the imperial palace provided a model for the ranking military and administrative households. Slave women populated the royal harem and sultans were the sons of slaves, their daughters

[36] Shaw, *History*, 1: ch. 2; Inalcik, *Ottoman Empire*, ch. 4.

marrying high-ranking slave officers and officials. Anyone who was part of the Ottoman state, from gardener to grand wazir, bore the title of "*kul*," all entered the system as slaves.[37]

The first half of the fifteenth century, between the death of Bayezid (1402) and the accession of Mehmed II (1444–1446, 1451–1488) was a period of disruption and dissension. Timur had overthrown Ottoman rule in Anatolia, and it was only slowly reestablished. Mehmed II's conquest of Constantinople in 1453 not only signified Ottoman recovery but also underlined the dynasty's ambition – to build an empire rivaling the Roman. Mehmed took two titles: Sovereign of the Two Lands (Rumelia or Southeastern Europe and Anatolia) and of the Two Seas (the Mediterranean and the Black). Although he expanded the new state's territories in both Anatolia and Eastern Europe, it was under Selim I (1512–1520) that the Ottomans became the most important Sunni state in the Islamic world. In 1514 Selim defeated Shah Ismail and his Turkish tribesmen at Chaldiran in eastern Anatolia. Although Selim couldn't hold Tabriz, he did expel the Safavids from Baghdad and Basra. In 1516 he defeated the Mamluk forces near Aleppo, adding Syria, Egypt, and the Arabian Peninsula to his fledgling empire. Gaining control of the two holy cities, Selim added, "Servant of Mecca and Medina," to Mehmed's titles.[38]

The Ottoman empire, like the Mughal and Safavid states, was agrarian-based. The bulk of imperial revenues came from rural taxes on crops, animals, and other produce. Whereas the Mughals and Ottomans measured the agricultural land of the central empire and drew up registers setting forth the average yield of each subdivision, neither had the fiscal structure or the administrative framework to centrally collect rural taxes and pay a corps of cavalry from the imperial treasury. As a result, the Ottomans, like the Mughals, paid their mounted men in grants of land revenue (called *timars*) – the *timar* system in the Ottoman empire was the rough equivalent of the *jagirdari* system in the Mughal.[39]

The *timar* was the smallest piece of assignable land. It consisted of a village or a group of villages and the surrounding fields. The sultan's share of the land tax was assigned to a cavalryman who lived in the village, maintained order, and joined the imperial forces whenever called upon. From the fourteenth through the sixteenth centuries, these men comprised the bulk of

[37] Colin Imber, *The Ottoman Empire, 1300–1650: The Structure of Power* (London: Palgrave Macmillan, 2002), 119ff; Inalcik, *Ottoman Empire*, ch. 11.
[38] Inalcik, *Ottoman Empire*, ch. 4; Shaw, *History*, 1: ch. 3.
[39] Imber, *Ottoman Empire*, 181–92, 234–61.

the imperial armies. The *timar* lands were mostly found in the older parts of the empire – Anatolia and the European provinces. After Selim's victory over the Mamluks in the early sixteenth century, Egypt and the North African provinces were not assigned to cavalrymen. They remitted a fixed annual sum in cash to the central treasury, most which went to pay the Janissaries.[40]

Unlike the Mughals or the Safavids, the Ottomans were a formidable naval power. After the conquest of Constantinople in 1453, the new capital was provisioned primarily by sea. With Selim's defeat of the Mamluks in the early sixteenth century control of the sea became even more important: ships full of Egyptian grain and cotton had to be protected and the pilgrimage routes from Africa and India had to be safeguarded. The Ottoman fleet was composed primarily of oared galleys, cannons being added in the late fifteenth century. Throughout the fifteenth and sixteenth centuries the Ottomans were the major naval power in the Mediterranean and Black Seas, occasionally even sending a fleet to the Indian Ocean – seventy-two ships were deployed against the Portuguese in 1538. Ottoman superiority at sea, however, ended in 1571 when Phillip II of Spain and the Venetians destroyed most of the Ottoman navy at the battle of Lepanto.[41]

In the older historiography the Ottoman empire was said to have reached the peak of its military, political, and economic power under Selim's son, Suleiman I (1520–66). Suleiman extended Ottoman rule in Eastern Europe by conquering Hungary (just failing to take Vienna in 1529). He defeated the Safavids and reconquered Baghdad and annexed most of North Africa. During Suleiman's reign also the Ottoman navy dominated the seas – the Mediterranean, the Black, the Red, and the Persian Gulf. Although in Europe Suleiman was known as the Magnificent (in honor of his military prowess and the splendor of his court), in his own dominions he was the Lawgiver. Although Quranic law regulated certain aspects of religious, family, and social life, many other matters were left to the decisions of individual sultans. Suleiman had all of these orders and opinions collected, collated, and, if necessary, revised. They were published in a single code, the Ottoman Laws. In Istanbul he increased the number of mosque schools, making a basic education available to all Muslim boys. For the talented, higher education could be pursued in one of the eight madrasas (colleges) of the capital. Suleiman also patronized the arts. Whereas painters, poets,

[40] Inalcik, *Ottoman Empire*, ch. 13.

[41] Imber, *Ottoman Empire*, ch. 9; Salih Ozbaran, "Ottoman Naval Policy in the South," in Metin Kunt and Christine Woodhead, eds., *Suleyman the Magnificent and His Age: The Ottoman Empire in the Early Modern World* (New York: Longman, 1995), 55–70.

jewelers, and goldsmiths produced important works, the greatest artistic achievements of his reign were probably architectural. Sinan, the chief imperial architect, was responsible for over three hundred monuments. His two masterpieces were the Suleimaniye Mosque Complex in Istanbul and the Selimiye Complex in Edirne (constructed during the reign of Suleiman's son Selim II).[42]

The Ottoman empire, like the Mughal, was a complex agglomeration of peoples and cultures, filled with different languages, religions, and ethnicities. Whereas the ruling elite was Muslim, it was not Turkish – as the sultans and many of the ranking officials were Christian converts or the descendants of Christian converts. In Anatolia, the ancient heartland of the dynasty, the people were Turks but in the former Mamluk territories Muslim Arabs predominated. And, of course, the European provinces held Greeks, Slavs, and Serbs – all Christian. In Istanbul there were a significant number of Jews. Within the overall framework of the Ottoman social order, the various religious and ethnic groups had a great deal of autonomy. Under the *millet* (community) system, a separate legal framework was established for each group. A leader resolved legal and social issues and served as the intermediary between the government and the community. As long as the various groups paid their taxes and were peaceful, the Ottoman authorities left them alone. The Muslim population itself was divided into two classes: the *askeri* (military) and the *reaya* (tax-paying). The *askeri* were mostly soldiers – either infantry or cavalry – but they also included other members of the Ottoman governing apparatus – administrators, courtiers, religious officials, teachers, and judges. All were paid by the state. The *reaya*, by contrast, were mostly peasant cultivators and, as the empire was agrarian-based, provided the bulk of imperial revenues.[43]

In the late sixteenth and early seventeenth centuries, after Suleiman and his immediate successors, the Ottomans faced a period of political unrest and military defeat. Two ruinous and ultimately unsuccessful wars tore at the military, economic, and political foundations of the state. In the east the war with the Safavids (1578–1590) added important territories but strained the capacities of the Janissaries and the provincial cavalry. And in the west the long and debilitating war with the Habsburgs (1593–1608), while ending inconclusively, revealed that the Austrians had become the military equals of the Ottomans. Meanwhile, Shah Abbas, after defeating the

[42] Shaw, *History*, 1: ch. 4; Inalcik, *Ottoman Empire*, ch. 5.

[43] Imber, *Ottoman Empire*, 230–38; Shaw, *History*, 1: ch. 6; Inalcik, *Ottoman Empire*, ch. 4.

Uzbeks, turned toward Anatolia. In the campaign (1603–1608) that followed he regained all of the territory that he had lost in the earlier war. At the same time, the slave system began to break down. The thirty-year period of almost continuous warfare bred an immediate need for more infantry and, as a result, many native born Muslims were enrolled in the ranks of the Janissaries. At the end of the sixteenth century, as the wars in the east and west wound down, the central administration demobilized a great many men. Banding together in groups of twenty-five to thirty, these unemployed soldiers (called *celali* in the sources) rocked Anatolia, Syria, and Iraq from circa 1580–1612. Hungry and experienced, they fought the imperial Janissaries to a standstill and forced many peasants to flee the countryside for the cities, causing a precipitous drop in agricultural production.[44]

In the older historiography the military defeats of the late sixteenth and early seventeenth centuries, along with the "*celali*" rebellions, were evidence of a precipitous decline from the glories of Suleiman's reign. The newer scholarship, however, has tended to reject this stark dichotomy and has begun to argue that the seventeenth century was more a period of change and transition than a time of decline and fall. The various crises brought about a profound transformation of the Ottoman state. Military defeats revealed that the *timar* calvary had become antiquated, increasingly ineffective in the new world of gunpowder technology. In order to recruit and train more infantry the imperial officials decided to increase the percentage of government revenues collected in cash, switching the tax collection status of more and more provinces from *timar* to tax-farming. At the same time under Ahmed I (1603–1617) the administrative system was reformed: a new law code was promulgated; the law of succession was changed, eliminating fratricide and opening the throne to collateral descendants; and the slave levy was slowly abandoned as more and more military and administrative recruits came from the free-born Muslim population. All of this led to a more professional organization that depended less and less on the energy and abilities of individual sultans, finally culminating in the rise to prominence of the Koprulu family of grand wazirs (1656–1703).[45]

[44] Imber, *Ottoman Empire*, 119–35; Shaw, *History*, 1: ch. 6; William J. Griswold, *The Great Anatolian Rebellion 1000–1020/1591–1611* (Berlin: Klaus Schwarz Verlag, 1983).

[45] Abou El Haj, *Formation of the Modern State: The Ottoman Empire, 16th to 18th Centuries* (Albany: State University of New York, 1991); Inalcik, *Ottoman Empire*, ch. 6; Suraiya Faroqhi, "Crisis and Change, 1590–1699," in Halal Inalcik and Donald Quataert, eds., *An Economic and Social History of the Ottoman Empire, 1500–1914* (Cambridge: Cambridge University Press, 1994).

Although the Mughal and Safavid empires were early modern entities entirely (born in the early sixteenth century and destroyed or substantially weakened by the early eighteenth century), the Ottoman empire was not. Founded in the late thirteenth century, it lasted until the early twentieth. The late eighteenth and early nineteenth centuries, however, were a time of military and economic weakness. Ottoman provinces in the Balkans were recaptured by Austria, and Egypt and Algeria became independent in all but name, eventually falling under the influence of Britain and France, respectively. Control of the core empire devolved from the Ottoman central government in Istanbul to local notables in the provinces. After a series of not very successful wars with the Russians, Selim III (1789–1807) began the process of military modernization. Reform and rejuvenation, spurred on by challenges from the European powers, continued through the Tanzimat (Reorganization) Period (1837–1876). In the late nineteenth century, nationalist movements erupted throughout the empire, and the Ottoman decision to enter World War I on the side of the Central Powers was catastrophic. At the Versailles Peace Treaty in 1919 the empire was dismembered, only Anatolia and a small slice of Europe remaining. When the Kemalist movement declared a republic in 1923, the career of the most successful Islamic state in history had come to an end.[46]

The political relationships among the three early modern empires were determined, to a significant degree, by the difficulties of distance. Whereas the empires differed dramatically in population, each controlled a substantial territory. Their capitals, however, were widely separated. It was 2,239 kilometers (1,391 miles) from Istanbul to Isfahan (as the crow flies), 2,466 kilometers (1,532 miles) from Isfahan to Shahjahanabad, and 4,556 kilometers (2,831 miles) from the Ottoman to the Mughal capital. As a result, for one state to mount a military campaign against another was a complex and extremely expensive proposition. In addition, because the Safavid empire, with its mountains and deserts, separated the other two, the primary political rivalries were, for the most part, between the Ottomans and Safavids, on the one hand, and the Safavids and Mughals, on the other.

From the early sixteenth century to the mid-seventeenth, the Ottomans and Safavids were locked into a contentious and intermittently bloody rivalry. Because both dynasties drew their early followers from the same ethnic group, the stakes were high. Turkish tribesmen constituted the bulk both of the Iranian Qizilbash (the Shiite members of the Safaviyya Sufi order) and the Ottoman *timar* cavalry. The Ottoman efforts to pacify and

[46] Shah, *Ottoman Empire*, vol. 2.

control the eastern Anatolian–northern Iran area (centered on the cities of Baghdad and Tabriz) were sporadic. Because the campaigning season was ordinarily limited to the three months following the fall harvest and as both sides often employed a scorched earth policy, the usual result of these battles was engagement followed by withdrawal – either the Ottomans from Tabriz or the Safavids from Baghdad. The issues, however, were real and the struggles, although short-lived, were often destructive.

By contrast, the political relationship between the Mughals and Safavids was much more peaceful. As with the Ottomans and Safavids, distance was an important factor. Isfahan and Shahjahanabad were separated not only by thousands of kilometers but also by the Hindu Kush and Safid mountains and by the winds and waves of the Arabian Sea. A brief look at the quarrels over Qandahar illustrates the point. Whereas the city, some three hundred kilometers south of Kabul, changed hands several times during the sixteenth and seventeenth centuries, the transfers of control were rarely bloody. Often, in fact, they were peaceful: a commander was bribed and defected or, faced with a superior force, offered a prudent surrender. For the Safavids Mughal India was a land of opportunity rather than a battlefield. Poets, painters, soldiers, administrators, and religious dissidents entered the subcontinent, finding refuge and fortune there. In the eyes of the seventeenth-century poet Saib Tabrizi, the principal enemies of the Safavids were the Uzbeks and Ottomans, not the Mughals.[47]

In the case of the Mughals and the Ottomans, no long-term, meaningful political relationship was possible. The enormous distance between the two capitals and the intervening presence of the Safavids meant that diplomatic missions, not to mention military confrontations, were extremely rare.[48] In fact, the only hint of an armed engagement was the appearance of the Ottoman navy in the Indian Ocean in the 1530s. Although a sporadic exchange of ambassadors between the two courts can be traced in the sources, the concrete results were negligible, and the intent seems to have been primarily symbolic, oriented toward impressing an internal audience rather than reaching any tangible political or economic agreements.

The only sustained interaction between the two was religious not political. Because the Ottomans controlled the Arabian Peninsula and the Mughals were Sunnis, the Indian pilgrims and the Meccan authorities

[47] Newman, *Safavid Iran*, 117.

[48] For a discussion, see Naimur Rahman Farooqi, *Mughal-Ottoman Relations: A Study of Political and Diplomatic Relations between Mughal India and the Ottoman Empire, 1556–1748* (Delhi: Idarah-i Adabiyat-i Delli, 1989).

had a complicated relationship that extended beyond the thirty days of the pilgrimage month. The issues were practical – gifts to Hajj officials and Indian overcrowding of the sacred sites – and the individuals involved were middle-ranking – Meccan governors and caravan ship captains. As a result, the disagreements never reached an intensity that would have provoked a military response.

Whereas the Ottomans were the superior military power, with access to the advanced gunpowder technology of early modern Europe, the Safavids played the leading role in the cultural sphere. In the early modern Islamic world, Arabic was the language of religion but Persian was the language of literature (poetry, history, geography) and, increasingly, of philosophy and science. It was also the *lingua franca* of the time, much like French in eighteenth-century Europe, and in Mughal India it was the language of court and state – records, documents, and orders. Although the Mughal emperors spoke Turkish or Urdu (Hindawi) within their extended households, Persian was the medium of communication among the members of the multicultural *mansabdari* system. Persian was also the language of history, theology, philosophy, and science, virtually all of the written material of the Mughal state was in Persian.

In the early modern Ottoman empire Persian did not play as central a role. At court the language was Ottoman Turkish and the Christian servitors who manned the higher reaches of the administrative and military hierarchies learned it as young boys. Although Ottoman Turkish was heavily influenced by Persian and, to a lesser extent, by Arabic, it quickly became the language of state – records, orders, and everyday conversation. Nevertheless, the ability to read and write Persian was highly prized. A knowledge of the Persian literary classics – the poetry of Saadi and Hafiz, the mystic verses of Rumi, and the histories of the Persian masters (especially the *Shahname* of Firdausi) was absolutely essential. Without mastery of this material, an educated man would be lost at the Topkapi court or in the upper reaches of the legal, religious, and administrative hierarchies. For example, in the last half of the sixteenth century the official Ottoman historian, even though he ordinarily wrote in Ottoman Turkish, bore the title "*Shehnameci*," that is, the *Shahname* writer.

If, in the early modern Islamic world, the Ottomans were dominant militarily and the Safavids culturally, then the Mughals were the preeminent economic power. Economic relationships among the three empires, although complex in detail, were in outline fairly simple. The basic movement was goods and commodities from east to west and precious metals (gold and silver) from west to east. Cloth and spices were sent from India to the Iranian markets, and Safavid silk was transported to the Ottoman ports of Aleppo, Izmir, and Istanbul. Ottoman silver (much of it from the new world) paid for the Safavid silk and passed directly into the hands of the Indian merchants who supplied the Iranians with textiles and spices. In addition, Mughal India also exported financial expertise: Indian bankers and moneychangers dominated the bazaars of Isfahan, Bandar Abbas, Tabriz, and Qazvin.

Although the three dynasties shared a common religion, ethnicity, and political and economy structure, their sources of legitimacy differed. Each based its authority, or right to rule, on a different set of beliefs and claims. After Selim's defeat of the Mamluks, the Ottomans asserted their claim to the caliphate. They were the protectors of the two sacred cities, Mecca and Medina, and the successors to the Rightly-Guided Deputies of the prophet.[49] The Mughals, like the Ottomans, were Sunnis, but their claim to legitimacy was based on their ancestry. Babur, the founder of the dynasty, was a direct descendant of Timur. The Safavids, by contrast, were Shiite and their authority, like the Ottomans, had a spiritual basis. Ismail claimed to be a descendant of the Seventh Imam, an incarnation of the Mahdi, and the Murshid (Master) of the Safaviyya Sufi order. The Safavids also claimed a share of the divine right accorded kings in the ancient, Iranian imperial tradition.[50]

[49] H. T. Karateke, "Legitimizing the Ottoman Sultanant," in H. T. Karateke, ed., *Legitimizing the Order: The Ottoman Rhetoric of State Power* (Leiden: Brill, 2005), 13–52.

[50] Roger Savory, "The Safavid State and Policy," *Iranian Studies* 7 (1974): 16.

2

Calendar

In all three empires the liturgical Hijra calendar was the principal method of defining and organizing the day. In none of the three, however, did its supremacy go unchallenged. The Safavid, Mughal, and Ottoman states had all adopted solar calendars of one sort or the other. Because each of these calendars – the Indic, the Zoroastrian, and the Julian – were quite different, each of the three temporal systems took on a distinctive color and complexity. And in each state it was the *munajjim*, the time expert, who was responsible for managing the difficulties thrown up by the use of different calendars.

SAFAVID EMPIRE

In the Safavid empire the solar Zoroastrian calendar was an important part of the temporal system. Although day began at sunrise for the Zoroastrians, the Safavids retained the Islamic definition (sundown) and, like the Mughals, divided the twenty-four-hour period into eight watches – four for the daylight hours and four for the nighttime.[1] For Shiite Muslims prayers were scheduled three times a day rather than five – at sunrise, noon, and sunset.[2] Some early Safavid scholars had argued that the Friday congregational prayer was illicit following the disappearance of the Twelfth Imam. However, after some debate, the clerical leaders decided that a Friday noon prayer, using the Shiite phrases, was acceptable.[3]

[1] Chardin, *Voyages*, 4: 395.

[2] Chardin, *Voyages*, 7: 10–12, 55–7; J. Thevenot, *Travels into the Levant* (London, 1686), 108.

[3] Chardin, *Voyages*, 7: 33; Andrew Newman, "Towards a Reconsideration of the Isfahan School of Philosophy: Shaykh Baha'i and the Role of the Safawid Ulama," *Studia Iranica* 15 (1986): 181.

The traditional Islamic calendar – the seven-day week and the twenty-nine or thirty-day lunar month – governed most religious, cultural, and governmental activities. However, because Iranian Muslims retained a vivid memory of their pre-Islamic past – the glories of Persepolis, the conquests of Cyrus and Darius, and the wars with Alexander and the Greeks – the solar Zoroastrian calendar never fell completely out of favor. It had been adopted by the Achaemenids (559–330 BCE) during the reign of the emperor Xerxes (486–465 BCE). Attributed to the founder of the Iranian religion (Zoroaster's dates are uncertain), it probably derived originally from the calendar of ancient Egypt, which had been incorporated into the Achaemenid empire in ca. 525 BCE.[4] The Zoroastrian calendar had 365 days divided into twelve months of thirty days – the extra five added at the end. The months were:

1. Farvardin;
2. Urdubihisht;
3. Khurdad;
4. Tir;
5. Murdad;
6. Shahriyar;
7. Mehr;
8. Aban;
9. Azar;
10. Dai;
11. Bahman; and
12. Isfand.

Unlike the Islamic, Jewish, Christian, and Indic months, however, the Zoroastrian were not divided into weeks. Rather, each day had its own name:

1. Ormuz;
2. Bahman;
3. Urdubihisht;
4. Shahriyar;
5. Isfandarmaz;
6. Khurdad;

[4] *Encyclopaedia of Islam*, 2d. ed., s.v. "Djalali"; Francois de Blois, "The Persian Calendar," *Iran: The Journal of Persian Studies* 34 (1966): 48–50.

7. Murdad;
8. Dai-ba-Azar;
9. Azar;
10. Aban;
11. Rash;
12. Mah;
13. Tir;
14. Gosh;
15. Khur;
16. Mihrgan;
17. Sarosh;
18. Rashn;
19. Farvardin;
20. Bahram;
21. Ram;
22. Bad;
23. Dai-ba-Din;
24. Din;
25. Arad;
26. Ashtad;
27. Asman;
28. Zamiyad;
29. Marisfand;
30. Aniran.

Under the Safavids, however, this complicated system was simplified. The day names were dropped, and, for most purposes, a simple numbering system was substituted. The solar calendar was used primarily to date the traditional pre-Islamic celebrations – in particular Nau Ruz, the first day of the year (1 Farvardin).[5] The other three solar eras of the period (the Yazdegird, the Jalali, and the Turkish Twelve-Year Animal) also used the revised Zoroastrian calendar.[6]

In Safavid Iran, as in the other two early modern states, the principal timekeeping devices were the sundial and the water clock. The sundial allowed the *munajjim* to determine both the times for prayer and the location of the qibla niche. In the late sixteenth century a household *munajjim* of Shah Abbas I constructed a sundial for the new congregational

[5] *Encyclopaedia of Islam*, 2d. ed., s.v. "Tarikh," 261–2.
[6] Ibid. 267.

mosque in Isfahan.[7] The water clock, however, was more accurate and reliable. In 1324 the Rukniya Madrasa in Yazd housed an elaborate water clock, and in 1472 another one was constructed in Kashan – a waqf underwrote both its construction and maintenance.[8]

Mechanical clocks, however, began to appear in Iran by the beginning of the sixteenth century. Muhammad Hafiz-i Isfahani, a *munajjim* and the premier horologer of the Safavid period, saw one in Herat in the late fifteenth century, and he constructed three himself: in Herat, Samarqand, and Kashan. Later, he authored a book on mechanical clocks – *Kitab Natijat al-Daulah* (1542–1543) – with designs, illustrations, and detailed instructions. In 1542 the Italian Michele Membre saw a large, elaborate tower clock in Tabriz. It featured clowns and revolving horsemen and showed the eclipses of the moon. Since Isfahani's book and Membre's visit occurred at roughly the same time, it is tempting to attribute the timepiece to the *munajjim*. Although European travelers of the next 150 years did not fail to notice the occasional Iranian clockmakers and repairmen, their testimony, for the most part, suggests that the most elaborate clocks and the most skilled clockmakers were European. Under Shah Abbas I an old Frenchman was the chief expert and when he could no longer work the Shah asked that another foreigner be sent. An Englishman named Fessy made the large clock above the Qaisariya Gateway in Isfahan but after his death it fell into disrepair. Another tower clock (designed by a European) was erected in the royal square during the reign of Shah Abbas II (1642–1666). It was dismantled during the Afghan invasion of 1722.[9]

In contrast to the three seasons of early modern India (hot, rainy, and cold), early modern Iran featured only two – hot and cold. Because of the temperature extremes a good part of the population – both rural and urban – regularly migrated in order to escape the worst of the weather. During the blistering summer months in the plains many people moved to cooler quarters (Turkish, *yailaq*; Persian, *sardsir*) in the mountains. During the chilly winter months, by contrast, they migrated to the warmer quarters (Turkish, *qishlaq*; Persian, *garmsir*) of the plains.[10] That these

[7] Siddiqi, "Construction of Clocks," 245–51; *Encyclopaedia Iranica*, s.v. "clocks;" Parviz Mahebbi, *Techniques and ressources en Iran du 7e au 19e siècle* (Paris: Bibliotheque Iranienne, 1996), ch. 9; Newman, "Isfahan School of Philosophy," 175.

[8] *Encyclopaedia Iranica*, s.v. "clocks"; Sayili, *Observatory in Islam*, 236.

[9] *Encyclopaedia Iranica*, s.v. "clocks"; Chardin, *Voyages*, 7: 354–5.

[10] Richard Tapper, "Tribes," in Peter Avery, Gavin Hambly, and Charles Melville, eds., *The Cambridge History of Iran, vol. 7: From Nadir Shah to the Islamic Republic* (Cambridge: Cambridge University Press, 1991), 524–5; Gene R. Garthwaite, *Khans*

migrations were central to medieval and early modern Iranian life is seen in the chronological framework of many historians – from the Il-Khanid period (1251–1335) onward the movements to and from summer and winter quarters constituted one of the most popular ways of locating and dating events.[11]

MUNAJJIM

In Safavid Iran, as in the Mughal and Ottoman empires, the *munajjim* was the time expert. Although the early modern sources rarely mention female astrologers, they must have been unexceptional. Bija Munajjima, a resident of late-fifteenth-century Herat, was a poet as well. She cast horoscopes, drew up almanacs, and translated dates from one era and calendar to another.[12]

In Iran the pre-Islamic astronomical/astrological tradition was an amalgam. In the earliest treatises three bodies of knowledge were integrated: the Zoroastrian, the Indic, and the Greek (Ptolemaic). The title of one early work, *Kitab-i Zaradost (Book of Zoroaster)*, suggested the ancient Iranian strand while the first complete Pahlavi astronomical treatise – the *Zik-i Shatro-Ayar or Zij-i Shahriyaran al-Shah or Zij-i Sha*hi (ca. 450) – revealed a heavy Indic influence. Two later treatises – the *Zij al-Arkhand* (ca. 556) and a treatise compiled during the reign of Yazdegird III (632–651) – contained large amounts of Greek material. During the Sassanid period (224–651), the Iranian populace routinely consulted their local astrologers. Sabur (son of the Sassanid ruler Ardashir I (r. 226–241), warned by a court astrologer of an upcoming danger, left the palace and entered (incognito) the service of a village headman until the threat had passed. And the prophet Mani (ca. 210–276) employed many astrological concepts in his religious and philosophical writing.[13]

Although the pre-Islamic astronomical/astrological system differed significantly from the Islamic – containing, for example, a great deal of material on Zoroastrian festivals and rituals, on the Zoroastrian era and

and Shahs: A History of the Bakhtiyari Tribe of Iran (Cambridge: Cambridge University Press, 1983), 20–21.

11 R. Stephen Humphreys, *Islamic History: A Framework for Inquiry* (Princeton, NJ: Princeton University Press, 1991), 19–20; Chardin, *Voyages*, 2: 286.

12 Maria Szuppe, "The Jewels of Wonder: Learned Ladies and Princess Politicians in the Provinces of Early Safavid Iran," in Gavin R. G. Hambly, ed., *Women in the Medieval Islamic World: Power, Patronage, and Piety*, (New York: St. Martin's Press, 1998), 329.

13 *Encyclopaedia of Islam*, 2d. ed., s.v. "Munadjdjim"; David Pingree, "Astronomy and Astrology in India and Iran," *ISIS* 54 (1963): 229–46.

calendar, and on Achaemenid theories about the world year – it did not long survive the arrival of the new religion. In Safavid Iran, in contrast to Mughal India, the local *munajjim* did not have to compete with different religious and cultural traditions. With the defeat of the Sassanids by the Umayyads in 637 and the slow conversion of the populace from Zoroastrianism to Islam, the Islamic temporal system slowly absorbed and replaced the Zoroastrian. In the first centuries of Islamic rule Arabic was the language of science, and the Iranian *munajjims* soon had access to a wide variety of Greek and Sanskrit treatises in Arabic translation. By the mid-eleventh century, furthermore, Iranian astronomers had fashioned a new synthesis – the language was Persian but the system was predominately Ptolemaic, with Sanskrit, Zoroastrian, Assyrian, and Syriac elements added.[14] For example, when the Buyid (945–1055) chief astrologer drew up the horoscope for the year (based on the ruler's horoscope), New Year's Day was 1 Farvardin (Nau Ruz) not 1 Muharram. By contrast, the Buyid ruler Abd al-Daulah (936–983), like other Islamic rulers, routinely celebrated his astrological birthday – the day when the sun reached the same point on the zodiac as when he was born.[15]

Because of the problems inherent in integrating the Zoroastrian (solar) and Hijra (lunar) temporal systems, the Iranians were in the forefront of observational astronomy in medieval Eurasia. They built the three earliest and most advanced observatories and produced the three most accurate and sophisticated astronomical treatises. The Seljuq ruler Malik Shah (r. 1072–1092) erected his observatory in Isfahan and his *munajjims* compiled the *Zij-i Malik Shahi*. The Ilkhanid ruler Hulaghu (r. 1256–1265) constructed his observatory at Maragha, near Tabriz, and Nasir al-Din Tusi (1201–1274) and his collaborators produced the *Zij-i Ilkhani*. Finally, the Timurid ruler Ulugh Beg (1394–1449) constructed an observatory in Samarqand and oversaw the composition of the *Zij-i Sultani* or *Gurgani*. After Ulugh Beg, however, the Iranian interest in observational astronomy seemed to die out. Although the first two Safavid rulers, Ismail and Tahmasp, considered building new observatories – Ismail wanted to renovate Maragha – nothing substantial came to pass.[16]

In addition to spurring the construction of observatories and the compilation of astronomical treatises, the Zoroastrian temporal tradition was

[14] *Encyclopaedia Iranica*, s.v. "Astronomy and Astrology."

[15] George Saliba, "The Role of the Astrologer in Medieval Islamic Society," *Bulletin d'etudes orientales* 44 (1992): 60.

[16] Sayili, *Observatory*, 166.

probably also responsible for the fact that the most popular solar eras of early modern Islam – the Yazdegird Era (epoch 632), the Jalali Era (epoch 1079), and the Turkish Twelve-Year Animal Era were of Iranian origin.

By the Safavid period the *munajjim* had become one of the most important men at court. In the administrative manuals the duties of the Chief Astrologer (Munajjim Bashi) included: determining the proper moment to begin important activities – battles, marriages, entries, or promotions; casting the emperor's horoscope at birth (a *kitab-i viladat* or nativity book),[17] on Nau Ruz, and on his birthday; recording and dating state orders; predicting eclipses, comets, equinoxes, and solstices; and providing equivalent dates across eras and calendars. He also hired and supervised the other imperial astrologers. On imperial seals and orders he was the Favored of the Emperor, the Ptolemy of the Age, and the King of Astrologers. In the cockpit of court intrigue the physician was his natural rival. The medical man's decisions, however, had greater consequences: "... if an astrologer makes a mistake," one practitioner told Chardin, "the sky reveals it; but when a physician makes one, ... the ground covers it."[18]

Because of the encyclopedic work of Jean Chardin, we know a great deal about the work of Safavid *munajjims*. The Frenchman devoted an entire chapter of his ten-volume work to astronomy and astrology and obtained a copy of the almanac for 1666. He provided a detailed description of its lists and tables. Since an almanac was by definition ephemeral (the information for a single year only), Chardin's copy was unique for early modern India and Iran. For India, there were examples from the eighteenth and nineteenth centuries (artistic creations for the elite), but nothing earlier.

Safavid astrologers used the *Zij-i Sultani* of Ulugh Beg to construct their almanacs.[19] Their principal instrument was the astrolabe and most practitioners made their own. As was the case in India, the almanacs for the rich and powerful were beautiful productions (painted and gilded with skillful calligraphy), while those for the man-in-the-street were cheap and hastily made.[20] The one for 1666 was divided into three parts: astrological, astronomical, and historical.

The astrological section began with praises of Allah (creator of the sky and the heavenly bodies), blessings on the prophets, and blessings on the

[17] Sergei Tourkin, "The Horoscope of Shah Tahmasp," in Serpi Bagci and Massumeh Farhad, eds., *Falnama: The Book of Omens* (London: Thames and Hudson, 2010), 327–31.

[18] Chardin, *Travels*, 9: 355.

[19] Ibid., 9: 324–5.

[20] Ibid., 9: 360–8.

inhabitants of the empire, especially the emperor. A universal horoscope, cast for the entire world, yielded a set of general predictions. For Iran the future was rosy, but for the Ottoman and Christian states the reverse was true. For India, China, and Central Asia (less serious political threats) more good than bad was expected. The generally positive overall prognosis for Iran was subdivided into a series of more specific predictions: for the imperial family health, economic prosperity, and political success; for the clerics a mixed outlook – the ulama would face problems and temptations, judges would not be entirely successful in detecting fraud and delivering justice, and students would achieve academic success. For generals, ministers, and magistrates positive achievements would be interrupted by difficulties and failures. Supervisors, secretaries, tax collectors, agents, and tax farmers would confront a difficult year of setbacks and losses. Peasants would find peace and fertility in the first half of the year but crop failure and default in the second half. Religious ascetics would be tempted to leave their simple lives and some would succumb. Women would experience occasional infertility and difficult deliveries. Artisans and laborers would find ease and abundance. For the natural world the almanac promised prosperity: health and abundant offspring for domestic animals and favorable harvests of wheat, cotton, fruits, cucumbers, oil, butter, dates, and sugar. A general cooling trend would mark the climate but that would not prevent the outbreak of fatal diseases. The political future of the empire as a whole was uncertain, with wars and rebellions in the offing.[21]

The second part of the 1666 almanac was astronomical. A collection of twenty-six tables presented a variety of information for each day of the year: the rising and setting of the sun and moon, their positions in the zodiac, the phases of the moon, the conjunctions of stars and planets, and solar and lunar eclipses. According to its astronomical character each day was labeled – white (good), black (bad), or indifferent. The last Wednesday of Safar was particularly inauspicious – Unhappy Wednesday.[22]

The final set of tables was chronological and historical. In Iran the astrologer was consulted just as often by the historian or the administrator as he was by the builder or general. The eight tables of the section included: the eras in use – Hijra, Yazdegird, Jalali, Turkish Twelve-Year Animal, Alexandrian, Nebucanezzarian, Rumi, and Elephant; the names of the days and months; the number of days in each year; and, most importantly, rules for converting dates from one era or calendar to another. There were

[21] Ibid., 9: 350–86.
[22] Ibid., 4: 446.

also tables listing the religious festivals of the Hijra calendar and the eight major festivals of the Iranian year.[23]

MUGHAL EMPIRE

In early modern India there were at least thirty Indic calendars (solar or lunisolar) in addition to the Islamic. And, in keeping with the multicultural, diverse character of Akbar's imperial order, the Indic calendars had a significant impact on the Mughal concept of time.

In the early modern Islamic world the fivefold division of the day (determined by the calls to prayer) had been supplemented by a more exact system. Babur, the first Mughal emperor, wrote:

> ... in our countries what is known as a ... *nychtemeron* [day-and-night] is divided into 24 parts, each called an hour [*saat*], and the hour is divided into 60 parts, each called a minute [*daqiqa*]. ...[24]

In India, by contrast, he found a different definition. The day began at sunrise not sundown, and the twenty-four hours were divided into eight *pahars* (watches), four for the day and four for the night. Each *pahar* contained sixty *gharis*, and each *ghari* was subdivided into sixty *pals*. Because the hours of sunlight and darkness fluctuated according to the seasons, the length of the *pahar* fluctuated as well, containing more *gharis* during the summer and fewer during the winter. This, however, was not unusual – in the early modern world the divisions of the *nychtemeron*, whatever they were called, were temporal (fluctuating according to the seasons) not equinoctial (of equal length).

Although Babur, who only lived in India for four years, employed the Islamic system of dividing the twenty-four-hour period, the other Mughal emperors, from Akbar onward, used the Indic method. In both the *Ain* and the *Akbar Nama*, Abu al-Fazl used *gharis* and *pals*.[25] The emperor Jahangir in his memoir (*Tuzuk-i Jahangir*) fixed the hour of his birthday celebration at three *pahars* and four *gharis*.[26] And the court historian described Shahjahan's typical day as consisting of two *gharis* for *darshan* (viewing), four to five *gharis* for general audiences, and four to five *gharis*

[23] Ibid., 9: 420–30.
[24] Zahir al-Din Muhammad Babur Padshah, *Babur Nama (Memoirs of Babur)*, trans. Annette Susannah Beveridge, trans. (New Delhi, 1970), 516.
[25] Abu al-Fazl, *Ain* 2: 13; Abu al-Fazl, *Akbar Nama* 1:54–5, 3: 372–3.
[26] Jahangir, *Tuzuk*, 1: 78.

for private audiences.[27] His court astrologers dated his birth at twelve *gharis* and three *pals* of a certain night.[28]

To keep track of the *pahars* and *gharis* the Indians used the water clock (*clepsydras*). Babur wrote:

> … a body of *gharialis* [timekeepers] is chosen and appointed in all considerable towns of Hindustan. They [hang] a broad brass [plate], … a *gharial* in a high place … [and] they have a vessel perforated at the bottom like an hour-cup and filling up in one *ghari* … when it fills the first time, they strike the gong once with their mallets; when a second time, twice and so on. … They announce the end of a watch (*pahar*) by several rapid blows of their mallets. … After this they pause; then strike once more, if the first day watch has ended, twice if the second, three times if the third, and four times if the fourth. After the fourth day watch, when the night-watches begin, this is gone through in the same way. It used to be the rule to beat the sign of a watch only when the watch ended; so that sleepers chancing to wake in the night and hearing the sound of a third or fourth *ghari*, would not know whether it was of the second or third night watch. I therefore ordered that at night or on a cloudy day the sign of the watch should be struck after that of the *ghari*, for example, after striking the third *ghari* of the first night watch, the *gharialis* were to pause and then strike the sign of the watch, in order to make it known that this third *ghari* was of the first night-watch. … It did very well; anyone happening to wake in the night and hear the gong, would known what *ghari* of what watch of night it was.[29]

Babur's system continued more and less unchanged throughout the period. Father Monserrate commented on the water clocks in Fathpur Sikri,[30] and John Fryer, who visited India in the 1670s, observed that the muezzins used them to determine the times of prayer: "… they have no Watches or Hour-Glasses but measure Time by the dropping of water out of a brass Bason … at the end of which Pores [*pahars*] the Priests Ascend their Steeples and are Monitors to them of their devotion …"[31] In fact, the water clock was still the principal method of domestic time management in the early nineteenth century.[32]

[27] Ibn Hasan, *The Central Structure of the Mughal Empire*; reprint ed. (New Delhi: Munshiram Manoharlal, 1970), 75–82.

[28] "Horoscope for Birth of Shahjahan," London, British Library Persian Manuscript Collection, Lansdown, 1245.

[29] *Babur Nama*, 516–17.

[30] *The Commentary of Father Monserrate, S.J., on His Journey to the Court of Akbar (1580–81)*, trans. J. S. Hoyland and S. N. Banerjee. (London: Macmillan, 1922), 211–12; Abu al-Fazl, *Ain* 3: 17–18.

[31] John Fryer, *A New Account of East India and Persia. Being Nine Years' Travels, 1672–81*, ed., William Crooke, 3 vols. (London: Hakluyt Society 1899), 2: 18.

[32] Mrs. Meer Hasan Ali, *Observations on the Mussulmauns of India: Descriptive of Their Manners, Customs, Habits, and Religious Opinions Made During a Twelve Years'*

In the Indic tradition the week was divided into seven days. Like the Babylonians and the Egyptians, the Indians named their days after the seven heavenly bodies: Sunday was Ravi-Var (Sun); Monday, Som-Var (Moon); Tuesday, Mangal-Var (Mars); Wednesday, Budh-Var (Mercury); Thursday, Guru-Var (Jupiter); Friday, Sukra-Var (Venus); and Saturday, Sani-Var (Saturn). The Mughals, however, while adopting the Indic method of dividing the day, employed the Persian variant of the Islamic day names. Thus Friday (sundown Thursday to sundown Friday) was Jama (Day of Assembly); Saturday was Shamba (Sabbath); Sunday, Yak-Shamba (Day One); Monday, Do-Shamba (Day Two); Tuesday, Si-Shamba (Day Three); Wednesday, Chahar-Shamba (Day Four); and Thursday, Panj-Shamba (Day Five).

Among the Mughals the Hijra definition of the month was widely employed: twelve months of twenty-nine or thirty days with the traditional names. For non-Muslims, however, the Indic "*tithi*" definition of the month generally prevailed. The "*tithi*," an astronomical term, was one-thirtieth of the time it took the moon, in its apparent motion, to travel around the ecliptic. The days of the month were divided into two fifteen-day halves. The first fifteen days ("*tithis*"), beginning with the new moon, were the bright half of the month, and the last fifteen, beginning with the full moon, were the dark half.[33]

The final natural calendrical division was the seasonal. For Babur and his Central Asian followers the Indian seasons were a surprise: "... whereas there are four seasons in those countries [i.e., outside India], there are three in Hindustan, namely, four months are summer; four are the rains; four are winter."[34] Over one hundred years later, Shahjahan's court historian also felt the need to carefully describe (the still strange?) seasonal characteristics of North India:

> This area ... has three seasons. The winter lasts from ... [late October to late January]; ... [December and January] are the coldest months; ... this season is the best in India, as during this time it is possible to travel and hunt to one's heart's content. The second, the summer, extends from [late February to late June.] ... [Late February] is the beginning of spring in India; it is very mild.... [late March to late April] also is mild; during these two months one can ride and run.... [late April

Residence in Their Immediate Society, 2 vols.; reprint ed. (Delhi: Idarah-i Adabiyat-i Delli, 1973), 1: 97–8.

33 Dewan Bahadur L. D. Swamikannu Pillar, *Indian Chronology* (Madras: Grant and Co., 1911), 10, 15; Raimundo Panikkar, "Time and History in the Tradition of India: Kala and Karma," in *At the Crossroads of Cultures and Time*, 80.

34 *Babur Nama*, 515.

to late May] also is not unpleasant, but it is not possible to exert oneself unless it is absolutely essential. In ... [late May to late June] the heat is at its height. The third is the rainy season, and this also if it rains, makes the air pleasant, otherwise it is hot like ... [late May]. ... [late July to late August] is the rainiest month, and the air is very pleasant while the rains last. During ... [late August–late September] also it rains, but not so heavily as in the previous month. ... [late September to late October] is the end of the rainy season; the rains during this period are particularly beneficial for the *kharif* [November harvest] and *rabi* [March harvest] crops.[35]

MUNAJJIM/JYOTISH

In the Indic temporal system, as in the Islamic, it was the astronomer/astrologer who was the time expert. According to an ancient Sanskrit text:

> The Veda exists for the purpose of accomplishing sacrifice; sacrifices are prescribed according to the order of the times; thus he who knows astronomy/astrology (*jyotisa*), which is the science that regulates time, knows the times of sacrifices.[36]

The Indic astronomical/astrological system, like the Islamic, was influenced by the Babylonians, the Egyptians, and the Greeks. Vedic astronomy (1000–500 BCE) was primitive – concerned with establishing the right times for sacrifices and determining the correct dates for equinoxes, solstices, and eclipses. During the Achaemenid Period (558–330 BCE) Indic astronomers employed Greek technical terms, Mesopotamian, Egyptian, and Iranian calendrical techniques, and Babylonian values for the *nychtemeron* (day-night). At this time also the Indians borrowed the Babylonian gnomon and water clock. Their early astronomical treatises (*siddhantas*) were mostly translations. In 149–50 CE an important Greek astrological treatise was translated into Sanskrit prose, followed in 269 CE by a poetic version entitled *Yavanajataka*. In the fourth century CE a Roman work was translated into Sanskrit under the title *Romakasiddhanta* (Roman Astronomical Treatise).[37]

Like his Islamic counterpart, the Indic *jyotish* derived the lion's share of his income from casting horoscopes. And, in order to construct his charts, he also had to consult an annual almanac (*panchangam*).[38] Although the almanacs (*panchangam/taqvim*) in both the Indic and Islamic systems

[35] Muhammad Salih Kanbo Lahauri, *Amal-i Salih*, ed. G. Yazdani, 3 vols. (Calcutta: Asiatic Society of Bengal, 1912–46), 3: 243–4.

[36] Panikkar, "Time and History in the Tradition of India," 81.

[37] North and Porter, *Astronomy*, 162–5; *A Concise History of Science in India*, eds. D. M. Bose, et al. (New Delhi: Indian National Science Academy, 1971), 130–4; Pingree, "Astronomy and Astrology."

[38] Pillar, *Indian Chronology*, 60.

performed approximately the same functions and included much of the same material, the astronomical treatises (*siddhanta/zij*) on which they were based differed significantly. Because of the radical redefinition of time introduced by the new religion, the Islamic *zij* contained a great deal of new material, tailored to the specific requirements of the new belief system: shadow lengths, solar or stellar time keeping tables, lunar visibility tables, longitudes and latitudes of towns and cities, tables listing the dates of important dynasties and festivals, and chronological conversion tables.

The Indic astronomical treatise, by contrast, reflected a completely different concept of time. It was more abstract and mathematical, less interested in practical observations, and the five principal *siddhantas* represented successive attempts to reach increasingly accurate solutions to different astronomical problems. The oldest, the *Surya Siddhanta*, appeared in about 400. Reflecting the influence of the Babylonians and Greeks, it replaced the outmoded Vedic theories and contained, among other things, the twelve signs of the zodiac, the correct length of the day and year, new data on planetary motions, and accurate dates for solstices, equinoxes, and eclipses. Although the identity of the original author is uncertain, the astronomers who wrote the other major astronomical treatises are well known. Aryabhata (ca. 476–550), the author of the *Aryabhata-Siddhanta*, assumed a heliocentric universe and calculated accurate approximations for several astronomical constants – the ratio between the rotations of the earth and the sun, for example. Brahmagupta (ca. 598–665) headed the astronomical observatory in Ujjain and wrote the *Brahma-Sphuta Siddhanta* (628) – in verse. This is the earliest known text, apart from certain Mayan works, to treat zero as a number in its own right. Brahmagupta was also the first person to use algebra to solve astronomical problems, and his *siddhanta* introduced Indian astronomy to the Islamic world. In 770 his treatise was translated into Arabic as *Zij al-Sindhind*. The third important Indian astronomer, Aryabhata II (920–1000), composed the *Maha-Siddhanta*.[39]

In keeping with Akbar's policy of "lasting reconciliation," the Mughal rulers employed both Islamic and Indic time specialists. In addition to Mulla Chand and Mir Fathullah Shirazi, Akbar also patronized the Indic astrologers Muni Sundar[40] and Nilakantha Ananta Chintamani,[41] and

[39] S. K. Chatterjee, *Indian Calendric System* (New Delhi: Publications Division, Government of India, 1998; Bose, *A Concise History of Science in India* (New Delhi: Indian National Science Academy, 1971), 81–125.

[40] D. N. Marshall, *Mughals in India: A Bibliographical Survey* (London: Asia Publishing House, 1967), 1320.

[41] Ibid., 1398.

the Jain Padmasundara.[42] Abd al-Rahim Khan-i Khanan, Akbar's multilingual great amir, composed an astrological treatise in mixed Persian and Sanskrit, utilizing material from both traditions.[43] Jahangir consulted Shrikrishna Daivajna,[44] Vishwanath Diwakar,[45] and Keshava Sharma, who held the title of Chief Astrologer (Jyotish-Rai).[46] The architect of Shahjahan's Taj Mahal, Ustad Ahmad, was an experienced astrologer. Having mastered the science of the stars, he had full command of Ptolemy's *Almagest*. Lutfallah, his son, was a poet, engineer, and astrologer – author of *Taqvim-i Lutf* (*Lutfi's Almanac*).[47] Mukramat Khan, an Iranian emigrant from Shiraz and governor of Delhi Province in 1641–1642, held the office of Superintendent of the Household Astrologers (Darugha-i Ahl-i Tanjum),[48] and Fazil Khan read the heavens for both Shahjahan and Aurangzeb.[49] Although by 1675 Aurangzeb had forbidden the preparation of almanacs, a *munajjim* calculated the right time for him to ascend the throne in 1658, and during his first decade he patronized the Indic experts Ishvar Das and Manirama Dikshita.[50]

Court astrologers, however, were only a small percentage of the total. Every village, town, and city had its resident experts – both Muslim and Hindu. Dargah Quli Khan described their activities in mid-eighteenth-century Shahjahanabad:

> Seated on wooden chairs are astrologers [*munajjims*] who speak eloquently on the relative importance of each day and month, the virtues of fasting in the month of Ramadan al-Mubarak, the performance of the pilgrimage in the holy month of Zu al-Hijja, and narrate the incidents from Rauzat al-Shahada [about the death of the Imam Husain during the month of Muharram]. … The audience give them their full attention. After making the listeners imbibe these details they make them part with the money in their pockets. … In their respective tents astrologers are occupied with their professions of unfolding the mysteries of the world to their

42 Ibid., 1442.
43 Ibid., 1518 (iv).
44 Ibid., 1711.
45 Ibid., 1834.
46 Ibid., 739.
47 M. Abdullah Chaghtai, "A Family of Great Mughal Architects," *Islamic Culture* 11 (1937): 202; Abd al-Qadir al-Badaoni, *Muntakhabu-T-Tawarikh*, trans. George S. A. Ranking, 3 vols.; reprint ed. (Delhi: Darah-I Adabiyat-I Delli, 1973), 3: 442.
48 Nawwab Samsam-ud-Daula Shah Nawaz Khan, *The Maathir-ul-Umara*, trans. H. Beveridge and Baini Prashad, 2 vols. (New Delhi: Janaki Prakashan, 1979), 2:1: 265–6.
49 Ibid., 1: 550–3.
50 Marshall, *Mughals in India*, 739; Rafat M. Bilgrami, *Religious and Quasi-Religious Departments of the Mughal Period 1156–1707* (Aligarh: Aligarh Muslim University, 1984), 194; Khan, *Maathir-ul-Umara*, 2:1, 45–6.

clients. People inquire from them about their good and ill fates and are extremely happy when they are told that good fortune awaits them . . .[51]

In order to construct a horoscope, the astrologer had to determine the exact moment of birth. In chapter two of Volume One of Abu al-Fazl's three-volume history of Akbar (*Akbar Nama*) the work of the court astrologers was recorded. They calculated the time (to the second) of Akbar's entry into the world, expressing the result in the various calendars and eras of the day. In chapter three the historian described the nativity chart drawn up by Mulla Chand, Humayun's court astrologer, using a Greek astrolabe and Ulugh Beg's *Zij-i Sultani*. Chapter four contained the horoscope produced by the emperor's chief Indic expert (Jotik Rai), and chapter five had the chart of the famous Safavid polymath Mir Fathullah Shirazi, employing Iranian techniques and Ptolemy's *Almagest*. Chapter six presented a detailed interpretation of the three previous horoscopes, while chapter seven gave the horoscope constructed by Maulana Alyas of Ardabil, employing Tusi's *Zij-i Ilkhani*. Chapter eight, in a summary, analyzed the differences among the various charts, both Islamic and Indic.

In addition to producing horoscopes, astrologers also played an important role in the decisions of daily life. Although Babur routinely consulted his household *munajjim* during the four years of his reign, his son, Humayun, displayed an interest that bordered on the obsessive. He assigned the planets to the days of the week and chose his clothes and rugs accordingly.[52] In 1547 he asked his court experts when the child Akbar should start school, and in 1555 he had them draw up plans for a new observatory. A site was chosen and instruments collected but his untimely death caused the project to collapse.[53] In 1590 Akbar's astrologers informed him that an impending eclipse heralded rain[54] but in 1596 they decided that a second eclipse foretold a drought.[55] In 1634 Shahjahan postponed a trip to Kashmir because it fell on the anniversary of his wife's death. The court astrologers chose a new departure date.[56] In 1639 the imperial astrologer determined the exact moment for laying the corner

[51] Dargah Quli Khan, *Muraqqa'-e-Delhi: The Mughal Capital in Muhammad Shah's Time*, trans. Chander Shekhar and Shama Mitra Cheny (Delhi: Deputy Publications, 1989), 21–2.

[52] Abu al-Fazl, *Akbar Nama*, 1: 643–4, 649–51.

[53] Ibid., 1: 519, 643–5.

[54] Ibid., 3: 878.

[55] Ibid., 3: 1064.

[56] Michael Brand and Glenn Lowry, *Fatehpur Sikri: A Sourcebook* (Cambridge, MA: MIT Press, 1985), 56.

stone of the new palace-fortress in Shahjahanabad.[57] A few years later Shahjahan approved a plan for the construction of a new observatory that, however, was never built – due to a shortage of funds.[58] Finally, during the early years of Aurangzeb's reign an administrative manual included a section on the lucky and unlucky days for important activities.[59]

Astrologers played a similar role in the households of high-ranking officeholders and provincial rulers. The mid-seventeenth-century household manual "*Bayaz-i Khushbui*" included a section on the proper times for bathing and traveling,[60] and, in the 1759 almanac drawn up for Nawab Mir Muhammad Jafar Khan of Bengal, the unlucky days for working, traveling, and planting were noted in red.[61]

To produce his horoscope the Mughal *munajjim* had to consult an almanac.[62] Drawing on the information contained in an astronomical treatise (*zij*), the annual almanac was a collection of lists, tables, and planetary coordinates. The versions for the rich and powerful were lavishly decorated while those found in the local bazaar were simple and cheap.

Although in the early modern Islamic world Ulugh Beg's *Zij-i Sultani* was the standard reference, in India two earlier astronomical treatises had been compiled. Mahmud bin Umar prepared *Zij-i Nasiri* for the Sultanate ruler Nasir al-Din Mahmud bin Shams al-din Iltutmish (r. 1246–1265). Only one copy of the work has survived, and although it has not been evaluated, it was most likely an update of Omar Khayyam's *Zij-i Malik Shahi*.[63] The second treatise, *Zij-i Jami Mahmud Shah Khilji*, completed in 1461–1462, was named after Mahmud Shah Khilji (1436–1469). It covered the usual topics and was heavily influenced by Tusi's *Zij-i Ilkhani*.[64] Under the Mughals three new *zijs* appeared. Mulla Chand, *munajjim* for both Humayun and Akbar, prepared for the latter *Tashil-i Ulugh Beg*.

57 Stephen P. Blake, *Shahjahanabad: The Sovereign City in Mughal India, 1639–1739* (Cambridge: Cambridge University Press, 1991), 29–30.

58 S. A. Khan Ghori, "Development of Zij Literature in India," in B. V. Subbarayappa, *The Tradition of Astronomy in India: Jyotihsastra* (New Delhi: Centre for Studies in Civilizations, 2008), 392

59 "Zawabit-i Alamgiri," London, British Library Persian Manuscript Collection, Or. 1641, fol. 184b.

60 "Bayaz-i Khushbui," London, British Library Persian Manuscript Collection, Ethe 2784, fol. 154a.

61 "An Almanac for the Year of the Hare 1172–73/1759," London, British Library Persian Manuscript Collection, Lansdown, 1245.

62 Abu al-Fazl, *Ain* 2: 12; W. Hunter, "Some Account of the Astronomical Labors of Jayasinha," *Asiatic Researches* 5 (1798): 178–80.

63 Ghori, "Development of Zij Literature in India," 392.

64 Ibid., 393–5.

A shorter and simplified version of *Zij-i Sultani*, it was partially revised for the Indian environment – dropping, for example, the section on the Chinese and Uighur animal calendars.[65] In 1629–1630, however, Mulla Farid Ibrahim Dihlavi presented a new, more ambitious treatise (*Zij-i Shahjahani*) to the newly crowned Shahjahan. The preeminent astronomer/astrologer of his day, Mulla Farid had studied under Mir Fathullah Shirazi and had spent most of his career in the employ of Abd al-Rahim Khan-i Khanan. The new *zij* was the inspiration of Asaf Khan, Shahjahan's wazir, who wanted Mulla Farid to create a new era for the new ruler – like the *Tarikh-i Jalali* for Malik Shah or the *Tarikh-i Ilahi* for Akbar. Because, however, Akbar's new era had aroused a great deal of opposition and because Mulla Farid was at this point an old man, the new treatise was mostly an updating (although extensive) of Ulugh Beg's *Zij-i Sultani*. Because it provided an expanded list of eras (including Akbar's *Ilahi* and the Indic *Samvat*), a more complete inventory of festivals and celebrations, and a corrected and revised set of astronomical/astrological tables, perhaps the court historian's claim – that the *Zij-i Shahjahani* rendered Ulugh Beg's treatise superfluous – was not a total exaggeration.[66]

About one hundred years later another *zij* appeared. Maharajah Jai Singh (1686–1743), Hindu ruler of the Rajput state of Amber and a high-ranking *mansabdar* under the Mughal emperor Muhammad Shah (1719–1748), had, like Ulugh Beg, a passionate interest in astronomy and astrology. He assembled a large library and gathered together a group of astronomers – Muslim and European as well as Indic. Although his ambition – to replace Ulugh Beg's treatise – was not new, the age was. By the early eighteenth century, the forefront of Eurasian astronomy had shifted from the Muslim East to the European West. Copernicus (1473–1543), Galileo (1564–1642), and Kepler (1571–1630) had replaced Ptolemy's geocentric cosmology with a heliocentric one, and the telescope had made all earlier tables and catalogues obsolete.

Jai Singh himself was not unaware of these advances. His library contained the works of two European astronomers – the *Tabulae Astronomicae* of the Frenchman P. de la Hire (1640–1718) and the *Historia Caelestis Britannica* of the Englishman John Flamsteed (1646–1720). In 1728 he dispatched the Jesuit Father Figueredo to Portugal to report on the state of

[65] Ibid., 395–6.

[66] It was divided into four parts and covered 419 manuscript pages. Mulla Farid Ibrahim Dihlavi, "*Karnama Sahibqiran Sani Zij-i Sultani*," London, British Library Persian Manuscript Collection, Or. 372. For a description, see Ghori, "Development of Zij Literature in India," 396–8.

European astronomy, and in 1734 he invited two priests (amateur astronomers?) from the Jesuit mission at Chandernagore to come to Jaipur. On their way, they visited two of his observatories. He also had two small telescopes at his disposal.

Nevertheless, the five observatories that Jai Singh erected in North India between 1724 and 1734 were profoundly conservative. For the Rajah a new observatory, one that would yield more accurate results, was not with one with a telescope – as had been the case in Europe since Galileo's first use of the instrument in 1610. Rather, it was the traditional design, only larger. Thus, in the new observatory constructed on the outskirts of the Mughal capital of Shahjahanabad in 1724 the guiding principle was "bigger is better." The three major instruments – the equinoctial dial, the double hemisphere, and the flat astrolabe – were all enlarged masonry copies of those constructed by Ulugh Beg in Samarqand. In addition to the one in the Mughal capital, Jai Singh erected four more observatories: a slightly larger one in his new capital of Jaipur (1734) and three smaller ones in Benares, Mathura, and Ujjain. All five bore the Sanskrit name Jantar Mantar or Instrument for Calculation.

The result of all this labor – the *Zij-i Jadid-i Muhammad Shahi* (*The New Astronomical Treatise of Muhammad Shah*) of 1728 – was underwhelming. Jagannath, Jai Sing's chief astronomer, was a Hindu and well-versed in the Indic astronomical/astrological literature – especially the *Surya Siddhanta*. And the Rajah had ordered him to learn Arabic and Persian in order to read the Islamic works. He translated Ptolemy's *Almagest* into Sanskrit – as *Samrat Siddhanta* – and had completely mastered Ulugh Beg's *Zij-i Sultani*. But the European works in the Rajah's library were mostly beyond him. The new treatise contained, like the *Zij-i Shahjahani*, a new era named after Muhammad Shah and more accurate figures for planetary locations and more exact predictions for heavenly phenomena – eclipses, conjunctions, and comets. Although mention was made of Kepler (1571–1630) and Galileo, the impact of their discoveries was limited because the context in which the Rajah and his men worked remained geocentric. As a result, the Rajah's new *zij*, the result of four or perhaps seven years of observations, while a major advance over Ulugh Beg's treatise, was well behind the work being done by astronomers in London, Paris, or Rome.[67]

[67] "*Zij-i Jadid-i Muhammad Shahi*," London, British Library Persian Manuscript Collection, Add. 14373; W. Hunter, "Astronomical Labors of Jayasinha," 177–86; G. R. Kaye, *The Astronomical Observatories of Jai Singh*; reprint ed. (New Delhi: Indological Book House, 1973). See also Ghori, "Development of Zij Literature in India," 398–403.

OTTOMAN EMPIRE

In the Ottoman empire the lunar Hijra calendar was the principal method of organizing the day as well. The day began at sundown, but instead of dividing the twenty-four hours into eight watches, as did the Mughals and the Safavids, the Ottomans employed two twelve-hour subdivisions – sunrise to sunset and sunset to sunrise. The first day of the week was Friday (Jama), and the other days carried the traditional names.[68]

The Ottomans paid a great deal of attention to the timing of the daily prayers. After the conquest of Istanbul a *muvakkithane* (timekeeper's office) was established in the courtyard of most local mosques. It held the usual astronomical instruments (quadrant, astrolabe, sextant, and octant) and timekeeping devices (sundial and water clock).[69]

In the late fifteenth and early sixteenth centuries, the principal timekeeping devices remained the sundial and the water clock. In Istanbul, for example, the water clock in the Beyazit Mosque (1501–1506) established the time for the entire city.[70] By the middle of the sixteenth century, however, Ottoman *muneccims* had become dissatisfied with the older timepieces and had turned their attention to mechanical clocks.

The story of this shift in timekeeping technology is best told through the career of the most famous and talented *muneccim* of the early modern Ottoman empire, Taqi al-Din b. Muhammad b. Maruf. Born in Damascus in 1525 to a family of Turkish descent, he was educated in Damascus and Cairo. In Cairo, Taqi al-Din studied theology, philosophy, and Islamic law in addition to astronomy and astrology. He served as a religious judge in a town near Cairo before moving to Istanbul, probably in the late 1540s or early 1550s, to pursue his astronomical/astrological career. Like most *muneccims*, Taqi al-Din probably supported himself in those early years by casting horoscopes, answering questions, choosing days and times, and constructing and interpreting almanacs (*takvims*). No ordinary practitioner, however, Taqi al-Din soon revealed his remarkable talent. Like the other early modern polymaths (Mir Fathullah Shirazi in India and Mir Ghiyath al-Din in Iran), he wrote a great deal – twenty-five works in Arabic on arithmetic, algebra, optics, astrolabes, observational instruments, sundials, and clocks. He penned a critique of Ptolemy and the early

68 E. Ihsanoglu, "Some Remarks on Ottoman Science and Its Relation with European Science and Technology up to the End of the 18th Century," in Ihsanoglu, *Science, Technology*, 56; Raphaela Lewis, *Everyday Life in Ottoman Turkey* (London: Batsford, 1971), 120.

69 Ihsanoglu, "Ottoman Science," 18.

70 Lewis, *Everyday Life*, 91–2.

astronomers and compiled two *zijs* of his own. In 1571 he was elevated to the office of *muneccimbashi* by Suleiman's successor, Selim II (1566–1574), and he continued in that office until his death in 1585. Despite his wide-ranging genius and prodigious output, Taqi al-Din is best known for two things: mechanical clocks and observatories.[71]

In his famous work "The Brightest Stars For the Construction of Mechanical Clocks (1565–66)," he wrote:

> Its aim [i.e. time-keeping] is to determine the times of prayer without observing the movements of the heavenly bodies and without using the quadrants ...[72]

The timekeeping devices of his day, however, were seriously deficient.

> ... sand clock. There is no profit in it. ... We have to control the vessels otherwise the result will be wrong. To use it for a long time requires detailed knowledge about this instrument. The result will be approximate.
>
> ... the *clepsydras* [water clock]. ... There is no profit in it also. The difficulty of its construction is more than its profit, ... it is not possible to carry it from one place to another.[73]

The first mechanical clocks arrived at the Ottoman court in the early sixteenth century from Holland, Hungary, France, and Germany. According to Ogier Ghiselin de Busbecq (1520–1592), Austrian ambassador to the court of Suleiman, they were not enthusiastically received:

> No nation in the world has shown greater readiness than the Turks to avail themselves of the useful inventions of foreigners, as is proved by their employment of cannons and mortars, and many other things invented by Christians. They cannot, however, be induced as yet to use printing, or to establish public clocks, because they think that the scriptures, that is, their sacred books would no longer be scriptures if they were printed, and that if public clocks were introduced, the authority of their muezzins and their ancient rites would be thereby impaired.[74]

Not paying close enough attention, modern scholars have taken Busbecq's words to mean that the Ottomans had no interest whatsoever in mechanical clocks. And it is true that tower or turret clocks (Busbecq's large public clocks) were extremely rare in the sixteenth and seventeenth centuries. But this probably had more to do with their expense and accuracy than with

[71] *Encyclopaedia of Islam*, 2d. ed s.v. Taqi al-Din; *Encyclopaedia of Islam*, 2d ed. s.v. "Miqat."

[72] Sevim Tekeli, *The Clocks in the Ottoman Empire in the 16th Century and Taqi al Din's "The Brightest Stars for the Construction of the Mechanical Clocks"* (Ankara: Ankara Universitesi Basimevi, 1966), 141–2.

[73] Ibid., 143.

[74] Ibid., 124.

any religious qualms. In fact, it was not until the late nineteenth century that these large constructions, symbolic of the Ottoman desire for modernization, began appearing in public squares across the empire.[75]

By contrast, Taqi al-Din's treatise, finished in the last year of Suleiman's reign, reveals a knowledgeable and discriminating interest in timekeeping technology. Although he listed five different mechanical clocks, not much space was given to the first three – the tower clock, the pocket watch, and the domestic clock. The tower clock was impractical, and the first known example of the pocket watch appeared in Germany in 1524, not turning up in England before 1580. Probably Taqi al-Din's mention here was more theoretical than actual, referring to a design rather than to an actual working model. Domestic clocks, to the contrary, began to appear in Istanbul palaces and mansions by about 1550, and at the end of the century had apparently become quite common. Taqi al-Din described two kinds: the larger, which stood on brackets or was affixed to the wall, and the smaller, found on tables.[76]

Because of his interest in astronomy, it was the last two types of mechanical clock – the astronomical and the observational – that dominated Taq al-Din's treatise. The problem with the European imports was their accuracy – a margin of error of twenty minutes a day or more.[77] The astronomical clock, which reproduced the movements of the seven heavenly bodies, was constructed according to the Ptolemaic theory of the universe, with the earth at the center. In Europe the most sophisticated examples were from the mid-sixteenth century and displayed the holy days of the Christian liturgical year. Taqi al-Din's clocks showed the faces of the moon, the days of the week, the relationships between the sun and the moon, the position of the sun in the zodiac, the azimuths, latitudes, and ascensions of certain stars, and the times of prayer.[78] He wrote:

> ... in 971 AH (1563–64) I was dictated to construct a clock which will show the times of prayer. I thought on this problem and made a dial which had the desired times. These are the times of glorifying God – dawn, Friday prayer, midday, afternoon, evening, [and] midnight prayer.[79]

[75] Jason Godwin, *Lords of the Horizon: A History of the Ottoman Empire* (New York: Henry Holt, 1999), 306.

[76] Tekeli, *Clocks in the Ottoman Empire*, 133–5.

[77] Ihsanoglu, "Some Remarks," 55–6.

[78] Tekeli, *Clocks in the Ottoman Empire*, 129–35, 175–81.

[79] Ibid., 184.

The last type, the observational clock, was a significant advance. A specialized astronomical time piece, its purpose was to establish precise times for the observation of the heavenly bodies, and it had three dials – for hours, minutes, and seconds. The most accurate timepiece of its day, it gave Taqi al-Din's *zij* the most precise values for solar, lunar, planetary, and stellar movements.

The Ottoman interest in mechanical clocks survived Taqi al-Din, and by the end of the sixteenth century a guild of clockmakers was found in the imperial palace. Twenty-four examples of their craft survive to this day.[80] European clockmakers, experiencing an increased demand for their wares, established a colony on the Galata Hill.

Although in the Ottoman empire the day and the week were defined and named according to the Islamic calendar, the month was a more complicated matter. In certain situations the months of the Julian calendar were employed instead of the Hijra. Adopted by Julius Caesar in 45 BCE, the Julian calendar contained 365.25 days. It replaced an earlier lunisolar calendar that required the periodic intercalation of an extra month. Under the Ottomans the Julian calendar or the Rumi Takvim (Roman Calendar) first appeared as part of the solar financial era (Maliye Takvim) of Bayezid I (1481–1512). The twelve Roman months were transliterated into Ottoman Turkish as: Yanar, Febrar, Mart, Abril, Mayis, Yunyus, Yulyus, Agustos, Septuris, Uhturis, Nuvuros, and Dekuris.[81] Unlike the Mughals or the Safavids, the Ottomans did not generally employ the solar calendar in any public text – history, order, or inscription. Its use was restricted, for the most part, to the fiscal and accounting offices.

MUNECCIM

In the Ottoman empire the *muneccim* was the time expert as well. He acquired his training as an apprentice: at the home of an accomplished practitioner, under one of the *muneccims* in the local *muvakkithane*, or, if talented, in the office of the *muneccimbashi* (chief imperial astronomer/astrologer). Like Taqi al-Din, the young apprentice mastered the works of Ptolemy, Nasir al-Din Tusi, and Ulugh Beg. He also studied the writings of two early Ottoman astronomers: Bursali Kadizade-i Rumi (d. 1440),

[80] Ihsanoglu, "Some Remarks," 55–6.

[81] L.T. Darling, *Revenue-Raising and Legitimation: Tax Collection and Financial Administration in the Ottoman Empire, 1560–1660* (Leiden: E.J. Brill, 1997), 135.

a director of Ulugh Beg's observatory and a contributor to the *Zij-i Sultani*, and Ali Kushcu (d. 1474), student of Kadizade and *muneccim* for Mehmed II. Ali Kushcu authored twelve books on mathematics and astronomy, including a commentary on the *Zij-i Sultani*.[82]

At the Ottoman court the *muneccimbashi* headed a large office. He and his subordinates performed the basic astronomical/astrological duties for the imperial family. They determined the auspicious dates and hours for important events – imperial accessions, wars, weddings, and the launching of ships. They also kept track of and interpreted unusual astronomical events – comets, earthquakes, and eclipses (both solar and lunar). For the Ottoman *muneccim*, as for his Mughal and Safavid counterpart, preparing the annual almanac (*takvim*) was an important task. The earliest Ottoman *takvims* were for the imperial family and were quite short. The first two, from 1444 and 1446, were produced for Murad II (1421–1451). Based on Ulugh Beg's treatise, they opened with a brief historical section: the Islamic prophets, the early caliphs, and the rulers of the Seljuq, Karaman, and Ottoman dynasties. An overall prediction for the year was followed by astronomical and astrological information for each day.[83]

Over time, Ottoman *takvims* became longer and more elaborate. Examples include those produced by Sheikh Wefa in the late fifteenth century and by an alim named Darendeli in the early sixteenth century – both for Istanbul.[84] By the mid-sixteenth century, the Ottoman *takvims* had assumed a standard format, quite similar to those produced in the Mughal and Safavid empires. There were twenty-four pages, one double page for each month. The right-hand columns listed the days of the week and their dates in the various calendars (Hijra, Yazdegird, and Rumi). At the far right were historical, religious, and cultural notes – the beginning of the shipping season and the anniversaries, feasts, and festivals of Christians, Muslims, and Jews. The left-hand columns noted the positions in the zodiac of the sun, moon, and five planets. At the far left were astrological predictions – the day would be good, bad, or indifferent.[85] In the 1593 almanac, for example, the ninth of Safar was a good day for inviting people to dinner, the twelfth was unfavorable for submitting petitions to the sultan, the sixteenth was unlucky for traveling, and the

[82] Ihsanoglu, "Ottoman Science," 21–3.

[83] L. Menage, "The Beginnings of Ottoman Historiography," in Bernard Lewis and P. M. Holt, eds., *Historians of the Middle East* (London: Oxford University Press, 1962), 171–2.

[84] *Encyclopaedia of Islam*, 2d. ed s.v. "Miqat."

[85] *Encyclopaedia of Islam*, 2d. ed s.v. "Taqvim."

eighteenth was inauspicious for buying horses.[86] At the top and bottom of each page unusual events were listed – conjunctions of planets (esp. Saturn and Jupiter), oppositions of the sun and moon, and eclipses. Each almanac began with the position of the seven heavenly bodies on Nau Ruz and the horoscope for the year.

By the late sixteenth century Taqi al-Din and the other Ottoman *muneccims* had become profoundly dissatisfied with the *Zij-i Sultani*. Almost 150 years old and badly out of date, the almanacs based on it had become increasingly inaccurate. Thus in 1575 Taqi al-Din submitted a petition to Murad III (1574–1595) calling for the construction of a new observatory and the compilation of a new *zij*. Although his petition was granted, it seems likely that the emperor's passion for astrology and the support of Sokullu Muhammad Pasha, the Grand Wazir, and Khwajah Saad al-Din, Murad's childhood tutor, were more important that any great interest in astronomical accuracy. Nevertheless, in awarding Taqi al-Din an additional grant for the new institution, Murad referred to the religious requirement of accurate timekeeping and to his pride at being the first Ottoman ruler to found an observatory. Completed in 1577, the new institution occupied a hill in the European section of the city, overlooking the Golden Horn and the Bosporus.

From Taqi al-Din's works and Ala al-Din Mansur's poetic account of Murad III's reign, a fairly clear picture emerges of the buildings, instruments, occupants, and activities of the new observatory. The chief *muneccim*'s ambition was not modest. He intended the new institution to rival the great observatories of the past: Tusi's in Maragha and Ulugh Beg's in Samarqand. It comprised two structures. The larger contained offices and apartments for the fifteen astronomers and a library – the books came from the estate of a recently deceased nobleman. The other building – the small observatory – contained the most advanced equipment of the day. A great deal is known about these instruments because Taqi al-Din described them in a pamphlet, "The Observational Instruments of the Emperor's Catalogue." The ten instruments fell into two groups – the traditional (armillary sphere, paralactic ruler, astrolabe, quadrants, and sextants) and the new, invented by Taqi al-Din himself (a sextant with a cord to determine the equinoxes, a wooden quadrant for measuring azimuth and elevation, and the observational clock). For the work of the observatory a large staff was required. Three men were assigned to each instrument: two manipulated the hardware and one wrote down the

[86] Lewis, *Everyday Life in Ottoman Turkey*, 120.

results. At this time the Istanbul Observatory was probably the most advanced in the world, superior in size and equipment to Tycho Brahe's contemporary structure at Uraniborg (1576).[87]

For Taqi al-Din the magnificent physical plant was not an end in itself. Rather, it was a preliminary step toward an even greater achievement – a full revision of Ulugh Beg's *Zij-i Sultani*. As we have seen, however, that was a lengthy and expensive ambition. Because of the distance from the earth of Jupiter and Saturn, a complete set of fresh observations would have taken thirty years. Unlike Omar Khayyam in the eleventh century, Taqi al-Din does not appear to have revealed this fact to his patron. From the evidence it appears that he and his men began preliminary observations in 1573 and worked for the next seven years – through the completion of the observatory in 1577 and for three very productive years afterward. From this labor two new *zijs* emerged. The first, the *Zij-i Sidrat Muntaha ... al-Falak al-Dawar*, employed a new method for determining the latitude and longitude of the stars using the planet Venus. New values for the sun's orbit were also calculated. Taqi al-Din was the first astronomer to use decimal points and with his observational clock he measured the right ascension of certain stars. His results were more accurate than those of either Tycho Brahe or Copernicus. His second *zij*, *Khidrat al-Durar wa Jaridat al-Fikar*, included new tables for the movements of the moon.[88]

The new observatory, however, the first of its kind in Ottoman Istanbul, suffered an unhappy fate. It opened in early 1577 and in November a comet flared across the Istanbul sky. Regarding this as a rare opportunity, the chief astronomer/astrologer wasted no time:

> The wise and sagacious man of learning, Taqi al-Din,
> The virtuous and illustrious man of wisdom,
> Worked for many nights without food or rest,
> To determine the implications of the appearance of this fiery body.[89]

Keeping in mind the first lesson of astrology – tell your audience what it wants to hear – Taqi al-Din arrived at court:

> Oh, world-swaying king!
> The candle of your pleasant society shall be resplendent.

[87] For Taqi al-Din's catalogue and *Zij*, see "Astronomical Instruments for the Royal Zij," and Zij-i Sidrat al-Muntaha al-Afkar," in Sayili, *Observatory*, 289–305; *Encyclopaedia of Islam*, 2d. ed s.v. "Taqi al-Din."

[88] *Encyclopaedia of Islam*, 2d. ed., s.v "Taqi al-Din."

[89] Sayili, *Observatory*, 291.

There are joyful tidings for you concerning the conquest of Persia,
For the foe is lying, with failing breath, upon the ground.[90]

Despite this auspicious beginning – a rare astronomical event coinciding with its inauguration – the new institution soon became a pawn in a power struggle at the highest levels of the Ottoman court. A contending faction led by the chief religious leader, the Sheikh al-Islam Qadizada, launched an attack on the party of the Grand Wazir. Qadizada declared that Taqi al-Din's interpretation had been wrong. The victory over the Safavids had been short-lived, a plague had devastated several cities, and more than one important man had died. Furthermore, observatories were sacrilegious (they pried into the secrets of nature) and often brought misfortune on their founder (Ulugh Beg was beheaded and his dynasty, the Il-Khanid, survived a mere seventy years).[91] In the end, the Sheikh al-Islam prevailed, and in January 1580, less than three years after completion, Murad ordered the observatory pulled down.

Before making his decision, Murad consulted Taqi al-Din. A veteran of court intrigue, realizing that he had been caught on the wrong side of a power struggle, the chief astronomer assured the sultan that his work was complete.

In the *zij* of Ulugh Bey
There were many doubtful points, oh exalted king;
Now through observations the tables have
Been corrected.[92]

Although Taqi al-Din knew enough to gracefully accept defeat, the destruction of the observatory when his revision of Ulugh Beg was far from finished must have been heartbreaking. He had five more years to live but they cannot have been happy.

One of the most important aspects of the Islamic redefinition of time was its division of the day. The institution of the five daily prayers bequeathed to early Muslims an intense interest in timekeeping. In the first centuries after Muhammad's death the principal devices were the heavenly bodies – the sun, moon, planets, and stars – and Islamic *munajjims* were preoccupied with tracking and timing their movements. Their solar, lunar, and stellar tables were the most accurate in medieval Eurasia, and their sundials and water clocks (witness Harun al-Rashid's gift to Charlemagne) were the most

[90] Ibid.
[91] Ibid., 291–2.
[92] Ibid., 293.

sophisticated. By the early modern period this obsession with exactly pinpointing the moment (of a birth, an eclipse, an equinox, a solstice, the rising of the moon, or the setting of the sun) had led Safavid and Ottoman astronomers to the mechanical clock.

The mechanical timepiece, invented in Europe in the early fourteenth century, had made its way east by the early decades of the sixteenth century. In Iran Muhammad Hafiz-i Isfahani concentrated primarily on showpieces – large tower clocks erected in the piazzas, bazaars, and gateways of big cities. Although he had planned and constructed three of these, his expertise and enthusiasm did not lead to any sustained interest among his countrymen. Rather, for the remainder of the period the principal experts appear to have been mostly European.

In the Ottoman empire the situation was different. Although Taqi al-Din's treatise followed Isfahani's by some thirty years, the Ottoman commitment to timekeeping technology was practical, technological, and long-lasting. In the sixteenth and seventeen centuries the Ottomans, unlike the Safavids, had very little interest in tower clocks – impressive architecturally but inaccurate and difficult to maintain. Rather, Taqi al-Din's motivation was scientific. For him, increasingly accurate clocks were a necessary preliminary to increasingly precise astronomical observations. And his results, more exact than those of Tycho Brahe at Uraniborg, suggest that he had in fact constructed the most accurate timepiece in the Eurasian world. More important, Taqi al-Din's initiative did not die stillborn. As *munaccimbashi*, he headed a large office and was able to pass on his interest and knowledge.

In the building, staffing, and equipping of observatories Islamic leadership continued through the end of the sixteenth century. To update the *Zij-i Sultani*, the goal of Islamic as well as European astronomers, two strategies were available: enlarging the old observatories and redesigning the ancient instruments or building new observatories and inventing new instruments. Here also Taqi al-Din led the way. He convinced Murad III to underwrite and staff a new institution and to provide long-term funding. The promise of this auspicious beginning, however, was unfulfilled, the whole enterprise falling victim to political rivalry and religious obscurantism.

The Indian chapter of the story, although different in detail, was equally unhappy and unproductive. In Mughal India a Hindu rajah with a passion for astronomy decided to construct an old-fashioned observatory and to update an Islamic astronomical treatise written in Persian. His chief assistant was an Indic astronomer who had mastered the Sanskrit *siddhantas*, learned Arabic and Persian, and had translated Ptolemy into Sanskrit. Although the Rajah and his lieutenant had access to small telescopes and the Latin treatises of French and British astronomers, they were largely unaffected. Their treatise, whose premise remained geocentric, was completed in the mid-eighteenth century but was no match for European works of the same period.

3

Ceremony

In the decades after the Prophet Muhammad's death, two ceremonies were widely celebrated. Tied to the defining practices of the emerging religion, these took place in the last months of the liturgical year. In the first three days of the tenth month (Shavval), the faithful rejoiced over the end of the month-long fast, and two months later on the tenth day of the pilgrimage month (Zu al-Hijja), believers in Mecca and around the world sacrificed an animal. In the following centuries two other celebrations made their appearance. The prophet's birthday, celebrated on the twelfth day of the third month (Rabi al-Avval), quickly became an important occasion, and from the late ninth century devout Muslims began to mourn the death of Husain, the prophet's grandson and the third Shiite Imam, during the first part of the first month (Muharram).

Although all three early modern empires inherited this common ceremonial tradition, each assembled a cycle of rituals and celebrations that was unique. Each constructed a ceremonial round that melded the Islamic and the indigenous, combining one or more of the four canonical (but often reconfigured) rituals with observances that were either adopted from the surrounding culture or invented. In all three the ritual systems that emerged were amalgams of the canonical, the local, and the created.

SAFAVID EMPIRE

According to travelers and contemporary historians, four rituals dominated the ceremonial calendar in Safavid Iran. Three were religious, located on the lunar Hijra calendar, and one was secular, dated according to the solar Jalali calendar.[1] The three religious festivals were Id-i Fitr, Id-i Qurban, and Ashura. The fourth festival, Nau Ruz, commemorated the beginning of the Jalali Era year. Of the four, only the one celebrating the end of the month-long fast was integrated into the Safavid temporal system relatively untouched. The other three ceremonies were substantially redesigned.

Id-i Fitr, celebrating the end of the Ramadan fast, could not begin until the new moon was seen. In 1624–1625, the Russian Kotov reported that the festivities continued for the entire night – houses and shops were decorated and outlined in lights, and the sounds of kettledrums, trumpets, and pipes accompanied the feasts and merrymaking.[2] Forty years later Chardin witnessed similar scenes of release and celebration.[3]

Id-i Qurban (Festival of Sacrifice) took place on the tenth day of the pilgrimage month (Zu al-Hijja). At least once during his lifetime every believer was commanded to visit Mecca during the last month of the liturgical year. The pilgrimage activities, scheduled for the seventh to the thirteenth of the month, included circumambulating the Kaba, drinking from the well of Zamzam, praying, reciting the Quran, stoning the devil, and sacrificing an animal. According to tradition, Muhammad had gathered the faithful at his home in Medina on the tenth day of the month and had brought forward two white rams. He sacrificed the first saying "O Lord, I sacrifice this for my whole people, all those who bear witness to thy unity and to my mission."[4] He sacrificed a second in the same manner. A ritual of unity, the sacrifice reenacted Abraham's near sacrifice of Ismail (as recounted in the Quran). For the millions of Muslims around the world who were not in Mecca, the sacrifice of an animal (a chicken, sheep, goat, or cow) was a declaration of solidarity and a reminder of God's love and favor.

Although there seems to have been a large public ceremony in Herat in 1469,[5] there is very little evidence of Id-i Qurban in Safavid Iran before the time of Shah Abbas I. What animals were sacrificed? By whom? Was the shah involved? Under Abbas, however, a new variant on the traditional ceremony emerged. From a three-day, mostly private affair among family,

[1] See Chardin, *Travels*, 6: 369–72; 7: 251–3; P. M. Kemp, ed. and trans., *Russian Travellers to India and Persia (1624–1798): Kotov, Yetremov, and Danibegov* (Delhi: Jiwan Prakashan, 1959), 32–3; Fryer, *A New Account*, 3: 138.

[2] Kemp, *Russian Travellers*, 26–7.

[3] Chardin, *Travels*, 7: 135–9, 256–8.

[4] M. E. Combs-Schilling, *Sacred Performances: Islam, Sexuality, and Sacrifice* (New York: Columbia University Press, 1989), 56–7; Grunebaum, *Muhammadan Festivals*, 15–36.

[5] Babak Rahimi, "The Rebound Theater State: The Politics of the Safavid Camel Sacrifice Rituals, 1598–1695," *Iranian Studies* 37 (2004): 459, fn. 27.

friends, and neighbors, Id-i Qurban became a public ceremony. Orchestrated by the shah and his court, the ritual was prolonged for more than a week and revolved around the public sacrifice of a new animal – the camel. Several traditions were said to justify this innovation. According to one: "Ibrahim then received in a dream the order to make a sacrifice to God. In the morning, he sacrificed a bullock and divided its flesh among the poor. In the night the voice again said: "God demands a more valuable offering. He killed a camel."[6] By contrast, Chardin was told that Abbas had sacrificed a camel because Abraham had ridden one to Mecca and had offered up fifty of the animals after Ismail was spared.[7]

In the seventeenth century the revised ceremony centered on the camel. An extravagant sacrifice, well beyond the resources of the ordinary believer, the camel was sacred, blessed in Mecca and dispatched to Isfahan (according to one report) or a perfect, unblemished specimen from the imperial stables (according to another). Whatever its origin, the animal was dressed in rich robes and covered with garlands of fragrant flowers. Its bit and bridle were of gold and silver. Blessed by the leading ulama, it was paraded through the city streets, admired, petted, and fed by the local inhabitants.

On 10 Zu al-Hijja the camel was led to the place of sacrifice (*qurban gah*) beyond the city walls. Although the logic of the ceremony dictated that the shah himself administer the fatal blow, a stand-in (mayor or ranking urban official) often performed the task. After the execution – witnessed by high-ranking officials, important ulama, rich merchants, and guild and quarter leaders – the camel was butchered and the pieces distributed, the head to the shah and the remainder to the urban populace. A citywide wave of sacrifices followed. The well-to-do offered a ram – LeBruyn reported a total of fifty thousand in the early eighteenth century – whereas the poor settled for a cock.[8]

The roughly contemporaneous Allawi Dynasty (1666–present) of early modern Morocco also transformed Id-i Qurban into a public event. In the imperial palace in the capital city of Fez the king sacrificed an animal (a ram rather than a camel) in a public ceremony called the Great Sacrifice.

[6] *Encyclopaedia of Islam*, 2d. ed., s.v. "Qurban."

[7] Chardin, *Travels*, 9: 14–15.

[8] Chardin, *Travels*, 9: 7–18; Thevenot, *Travels*, 106–8; Cornelius Le Bruyn, *Travels into Muscovy, Persia*, 2 vols. (London: A. Bettsworth, et al., 1737), 1: 190; John Baptista Tavernier, *The Six Voyages of John Baptista Tavernier*, trans. J. P. London, n.p., 1678), 163; Fryer, *A New Account*, 3: 141; Raphael Du Mans, *Estate de la Perse en 1660*, ed., Francis Richard, 2 vols. (Paris: Societe d' Histoire de l'Orient, 1995), 73–5; Rahimi, "Rebound Theater State," 451–78; Jean Calmard, "Shii rituals and Power II," 139–90.

Because the Allawi rulers claimed direct lineal descent from the prophet Muhammad, the ritual was a symbolic reminder of the ruler's pedigree and of his mediating role between God and nation.[9]

In the Safavid empire, however, the most important religious ritual was Ashura. Celebrated during the first ten days of the liturgical year, it dominated the ceremonial calendar. In its earliest form Ashura (lit., Tenth) was not connected to Shiite mourning observances at all. According to tradition Muhammad immigrated to Medina in September of 622 and decided to observe the Yom Kippur fast (10 Tishri), which occurred in the fall during the first month of the Jewish lunisolar calendar. In that year the Islamic Ashura (10 Muharram) and the Jewish Ashura were said to have fallen on the same day. In any event, by the second year of the Hijra Ramadan had become the month of fasting, and the Ashura fast had become far less important.[10]

Some fifty years later Ashura was completely transformed. In 680 the Umayyad ruler Muawiya (661–680) died and was succeeded by his son Yazid. The followers of Ali, Muhammad's son-in-law and the first Shiite Imam, opposed Yazid's elevation and urged Husain, Ali's son and the prophet's grandson, to mount a revolt. In the summer of that year on the plains of Kerbala sixty miles southwest of Baghdad Husain's small party was surrounded by the Umayyad troops. All of his men were killed; the women and children were taken captive.

The death of the prophet's grandson triggered an immediate reaction. When the bedraggled captives (Husain's head impaled on a long pole) were led through the streets of nearby Kufa, the people began to weep and beat their breasts. In Medina the martyr's only surviving son (a baby at Kerbala) was proclaimed the fourth Imam and his home was transformed into the center of a growing cult. While the Umayyads tried to suppress the Ashura ceremonies and the opposition they engendered, the Abbasids (750–1258) patronized and encouraged them. By the end of the tenth century Husain's martyrdom had become a favorite topic for professional storytellers. In 962 the Buyid (945–1045) ruler of Baghdad declared Ashura an official day of mourning.[11]

Although Ashura was commemorated in the Islamic world before the early modern period, it was under the Shiite dynasty of the Safavids, and

[9] Combs-Schilling, *Sacred Performance*, 10–11, 222–32.

[10] *Encyclopaedia of Islam*, 2d. ed., s.v. "Ashura"; Yitzhak Nakashi, "An Attempt to Trace the Origin of the Rituals of Ashura," *Die Welt des Islams* (Nov. 1993): 161–81.

[11] Grunebaum, *Muhammadan Festivals*, 85–9.

especially during the reign of Shah Abbas I, that mourning activities became widely popular. Although Shah Ismail, the founder of the dynasty, had headed a Shiite Sufi order and had written poems in praise of Ali and Husain, there are no accounts of Ashura commemorations during his reign.[12] Under Tahmasp, Ismail's successor, Shiites on the island of Hormuz organized mourning rituals, and in Tabriz in 1540 Michelle Membre witnessed a ten-day ceremony with processions, self-mutilation, and fighting. In the central mosque the women wept loudly during a sermon on Husain's sufferings. There is no evidence, however, of imperial patronage or participation.

For Shah Abbas, by contrast, the promotion of Ashura became an important part of his effort to spread Imami Shiism. Under the shah and his successors the period of mourning grew from one or two days to ten. While the activities themselves changed over time, the basic theme was set: lamentations for Husain and the other martyrs and a release from feelings of persecution, betrayal, and loss. In their re-creation of Husain's ordeal the rituals became increasingly loud, violent, and bloody. The earliest mourners wore black clothes. Their faces were painted black, and they pounded stones against their chests. They carried long sticks and chanted the names of the martyrs. Self-inflicted wounds sent streams of blood coursing down their cheeks. Later, the men carried bows, arrows, and shields and, dressed like the Karbela martyrs, followed a riderless horse with a turban on its saddle horn. Some mourners hoisted tall standards and carried coffins displaying the effigies of the martyrs. Inevitably, the charged atmosphere found an outlet in ancient rivalries and animosities, and periodic outbreaks of violence – sticks and stones the usual weapons – were common. The ten-day ritual became so popular that Abbas and his successors began to mark the occasion even when they were out of the city – in camp for a campaign or hunt.

The development of the Ashura ritual did not end with Abbas. During the seventeenth century the processions became longer and more elaborate. After winding through the streets of the capital, the mourners ended up in the central piazza of Isfahan where the Shah and his court witnessed the spectacle from the balcony of the Ali Qapu (the gateway to the imperial palace). In the late 1660s Chardin observed the entire ten days – processions of naked, bleeding men, costumed mourners, decorated horses, and cenotaphs for Husain. In 1704 Cornelius de Bruin described floats (wheeled

[12] Mayel Baktash, "Ta`ziyeh and Its Philosophy," in Peter J. Chelkowski, ed., *Ta`ziyeh: Ritual and Drama in Iran* (New York: New York University Press, 1979), 103.

platforms) displaying bloodstained sand, beheaded or mutilated bodies of straw, and severed arms and legs. *Nakhls* (date palm trees) represented Husain's bier. In the late seventeenth century also mourning assemblies began to appear. Dramatic recitations of the lives, deeds, and sufferings of Husain and his family, these were narrated by professional storytellers who sat on raised platforms in bazaars, tents, or mosques. The usual text was the *Rauza al-Shuhada* (Garden of the Martyrs) of Husain Vaiz Kashshifi (d. 1504–1505). This new form of mourning took place in a new structure – a *takiya or husainiyya*. These assemblies were the precursors of the famous passion plays (*taziya*) of the nineteenth century.[13]

The last of the four major Safavid rituals was quite different from the other three. Nau Ruz (New Day) was a solar not a lunar celebration and, under the Safavids, it was a secular rather than a religious ritual. A New Year's Day celebration, its exact origin is obscure. Writing in circa 1000, passed on the traditions he had collected: Nau Ruz celebrated the day Solomon regained his throne or the day the legendary Iranian king Jamshid reformed the ancient religion and built a new temple. In the legendary past its date was the summer solstice (21 June).[14]

Under the Achaemenid Dynasty (550–350 BCE), by contrast, Nau Ruz was celebrated on another date of solar significance – 21 March, the Vernal Equinox. At this point it was a religious festival, associated with Zoroastrian beliefs and practices. Although the exact dates of the prophet Zoroaster are uncertain – fifth or sixth century BCE perhaps – Zoroastrianism eventually became the national religion of the Achaemenids. By the latter part of the period, an annual cycle of feasts (the Gahambars) had become established, of which Nau Ruz was the most important. In 487 BCE Darius I (522–486 BCE) mounted a magnificent celebration in Persepolis.[15]

For the Zoroastrian Sassanids (224–651), as for the Shiite Safavids, Nau Ruz was an imperial ritual. At the Sassanid court, according to al-Biruni, it was a six-day affair. In successive audiences the ranking nobles and officials tendered congratulations and gifts. On the sixth day the emperor celebrated with his family. Omar Khayyam in his Nau Ruz Nama (New Year's Day Chronicle) wrote: "... on the first day of the New Year the King's first visitor was the high Priest of the Zoroastrians, who brought ... as gifts a golden

[13] For a complete discussion see Jean Calmard, "Shii Rituals and Power II," 139–90. See also Kemp, *Russian Travellers*, 31–2 for 1624–25; Thevenot, *Travels*, 107–08 for 1665; Chardin, *Travels*, 9: 46–65 for a detailed description of the great festival of 1667.

[14] Abu Rayhan al-Biruni, *Chronology of Ancient Nations*, trans. C. Eduard Sachau (London: Oriental Translations Fund, 1879), 199–200.

[15] Mary Boyce, "On the Calendar of the Zoroastrian Feasts, *BSOAS* 23 (1970): 538.

goblet full of wine, a ring, some gold coins, ... [He said] "O Majesty, on this feast of the Equinox, first day of the first month of the year. ... Live long in praise, ... seeing that thou has freely chosen God and the Faith of the Ancient Ones drink immortality from the Cup of Jamshid ..."[16]

With the coming of Islam and the imposition of the Hijra calendar and era, dating and celebrating Nau Ruz became part of the larger chronological problem: How to administer an agrarian state without a solar era and calendar? The Abbasid solution – the Kharaji Fiscal Era – was inadequate, and, with no regular intercalation, the date of the New Year's festival continued to regress. Nevertheless, under both the Umayyads and the Abbasids the celebration of the first day of spring (now an entirely secular affair) continued both at court and in the countryside. In the earlier dynasty gift-giving, feasting, and the sprinkling of water were common,[17] and under the Abbasids al-Mutawakkil (847–61) organized a lavish celebration, minting five million small coins and showering the assembled multitude.[18] Not until 1079, however, was the dating problem solved. The new era (the Jalali) established by Omar Khayyam and his colleagues included a new calendar. Nau Ruz was returned to 21 March, and the insertion of an extra day every four years ensured that the ritual would always coincide with the arrival of spring.

Despite the pivotal role of Shiism in Safavid ideology, Nau Ruz continued to be an important ritual – according to some travelers, the most important. Whereas the coming of spring was celebrated in the early sixteenth century under both Ismail and Tahmasp, most of the evidence comes from the reigns of Shah Abbas and his successors. Murshid Quli Khan, Abbas's tutor and regent, convinced the young ruler that it was symbolically important for him to publicly celebrate his first Nau Ruz (21 March 1588) in Qazvin (the Safavid capital). Two years later, having embarked on a comprehensive program of reform and having shifted his capital to Isfahan, Abbas made sure that the New Year (21 March 1590) was celebrated in style. In the bazaars the shops were decorated, and in the streets and public squares the people, wearing new clothes, threw red eggs at one another. The festivities lasted for three days.[19]

By the mid-seventeenth century the celebration had become longer. In the 1660s, according to Chardin, the ritual was called Nau Ruz-i Sultani

16 al-Biruni, *Chronology*, 203–04; Boyce, "Zoroastrian Feasts," 517–18, 537.

17 Aziz al-Azmeh, *Muslim Kingship: Power and the Sacred in Muslim, Christian, and Pagan Polities* (London: I. B. Tauris, 1997), 69.

18 Grunebaum, *Muhammadan Festivals*, 54–5.

19 Kemp, *Russian Travellers*, 26–7.

(the Imperial New Year) to distinguish it from the liturgical New Year (1 Muharram). Although the lunar New Year was not ordinarily celebrated in the Islamic world at large, in Safavid Iran it was completely ignored, overshadowed by the extravagant and emotional activities of Ashura (1–10 Muharram). At the appointed hour the chief imperial astrologer came to the palace and, after consulting his astrolabe, announced the precise moment of the equinox. Cymbals, drums, and trumpets blared the news to the city. At court the ceremony lasted for eight days. On the first day the shah held a public audience, receiving the congratulations and presents of the general populace. The following six days were devoted to the ranking men of the realm: scholars and astrologers, ulama, judges, high-ranking officials, merchants, and the imperial family. All brought valuable presents – golden eggs, jewels, and silks. Gold coins were expected from officials at court and in the provinces. To advance in the Safavid hierarchy the Nau Ruz offering had to be substantial, and the gifts themselves constituted a significant part of the shah's household income.[20] During the audience the shah offered the great men congratulations and ceremonial robes (*khilats*). The rich and powerful held their own assemblies, and the ordinary populace in new clothes exchanged visits and gifts.[21]

Nevertheless, the Safavid strategy of promoting Shiism began to gradually diminish the importance of Nau Ruz. In 1611, when the Vernal Equinox fell during the first half of Muharram, Abbas postponed the New Year festivities until after the tenth of the month. And, by the 1660s the ritual had been given a Shiite interpretation. It was said to also commemorate Muhammad's anointing of Ali as his successor. Chardin quoted a popular verse:

> The spring time rises with a tulip in the hand,
> ... To make a scattering of the drops of the sun
> On the tomb of the king who is in Najaf [i.e., Ali].[22]

Although the four major rituals of the Safavid year were inherited, rooted in either the Islamic or the Iranian past, only one – Id-i Fitr – entered the ceremonial round untouched. For the other three, a substantial reworking of traditional themes and activities occurred. The two Islamic ceremonies – Ashura and Id-i Qurban – were celebrated in contiguous months. Id-i Qurban took place on the tenth day of the last month of the Hijra year

20 Rudi Matthee, "Between Venice and Surat: The Trade in Gold in Late Safavid Iran," *Modern Asian Studies* 34 (2000): 235.

21 Chardin, *Travels*, 2: 249–73.

22 Ibid., 2: 271.

(Zu al-Hijja), whereas twenty days later the ten-day festival of Ashura began on the first day of the first month (Muharram). Even though the two ceremonies regressed against the seasons, never occurring at the same point in the annual cycle, they remained in close proximity. Although both were reorganized by Shah Abbas, it was the quintessentially Shiite Ashura that came to dominate the Safavid year. Not only was it expanded and enhanced, but it also influenced the celebration of Id-i Qurban. Although the worldwide reenactment of Abraham's near sacrifice of Ismail had no particular Shiite connection, in Abbas's redesign a new wrinkle was added. For both the Mughal and the Allawi dynasties, the ruler's sacrifice of a ram or a camel inaugurated an empirewide cycle of private sacrifices. Under the Safavids, by contrast, Id-i Qurban was transformed into a huge public ceremony in which the bloody pieces of the sacrificed camel were distributed throughout the city. This new element – the public display of a bloody victim – bore the unmistakable imprint of the upcoming Ashura ritual.

Although the Safavids added a Shiite ingredient to their Nau Ruz celebration, the ceremony was intended, for the most part, to underscore the dynasty's claim to the pre-Islamic Iranian tradition of divine kingship.[23] Nau Ruz commemorated the original creation of the world, and it was the ruler, according to al-Biruni, who presided over the process, announcing: "Here is a new day of a new month of a new year. What time has worn out must be renewed."[24] In reenacting this ritual, the Safavid Shah proclaimed himself the possessor of the divine aura and semidivine status of the kings of ancient Iran.[25]

MUGHAL EMPIRE

In the Mughal empire, in contrast to the Safavid and Ottoman empires, there were no national festivals. Instead there were many different celebrations, each keyed to one of the several calendars and eras. Different days were declared auspicious and commemorated according to religion (Hindu, Buddhist, Jain, Zoroastrian, or Muslim), sect (Sunni or Shiite Muslim,

[23] One scholar has argued that the Safavid empire drew on three sources of legitimacy: the pre-Islamic Iranian tradition of divine kingship, the claim of the Safavid shahs to be representatives of the Mahdi (the Twelfth or Hidden Imam), and their customary role as leader of the Safaviyya Sufi order. Roger Savory, *Iran Under the Safavids* (Cambridge, UK: Cambridge University Press, 1980), 1–3.

[24] Quoted in Mircea Eliade, *The Sacred and the Profane: The Nature of Religion*, trans. Willard T. Trask (New York: Harcourt Brace and Co., 1959), 78–9.

[25] Ibid.

Vaisnavite or Shaivite Hindu), ethnicity (Iranian, Afghan, or Turani), gender (male or female), region (Bengal or Maharashtra), class (rich or poor), or location (rural or urban). Despite this complexity, however, the ceremonial round in Mughal India came to be dominated by three cultural traditions – the Islamic, the Indic, and the Zoroastrian. In addition, Akbar created a new imperial ceremony of integration and inclusion.

Islamic (Lunar)

The Mughal court led the commemoration of the major Islamic rituals. Id al-Fitr, marking the end of the month-long Ramadan fast, was a three-day celebration. On the morning of the first day the faithful gathered in an Id-Gah (Place for Celebration) or Namaz-Gah (Place for Prayer) where the Imam led the prayers and delivered a brief address. Afterward, dressed in new clothes, they gave alms to the poor, visited friends, exchanged sweets, and set off fireworks.[26] Under Aurangzeb the celebration was briefly enlarged. Commemorating both the end of the fast and, for a time, his accession to the throne, the Id was celebrated for ten days instead of three.[27]

On Id-i Qurban the pious gathered for prayers in the local congregational mosque or Id-Gah. The defining act of the celebration was the sacrifice of an animal – a chicken, sheep, goat, or camel, depending on the wealth of the family. Part of the animal was given to the poor, part to relatives and friends, and the rest was eaten by the family.[28] Humayun celebrated the ritual in the early part of his reign,[29] and in 1556 the fourteen-year-old Akbar witnessed the sacrifice but, because of his age and inexperience, was not an active participant.[30] In later years, however, Akbar's policy of "lasting reconciliation" – its incorporation of the Indic opposition to meat eating and violence – led him away from the ceremony. In 1611, for example, Jahangir did not sacrifice an animal on the prescribed day because it was Thursday, his birthday. (The members of Akbar's Tauhid-i Illahi abstained from

[26] *Islam in India or Qanun-I Islam of Jafar Sharif*, trans. G. A. Herklots (Oxford: Oxford University Press, 1921), 144–5; P. N. Chopra, *Life and Letters under the Mughals* (New Delhi: Ashajanak Publications, 1976), 94–5; E. Dennison Ross, *An Alphabetical List of the Feasts and Holidays of the Hindus and Muhammadans* (New Delhi: Imperial Record Department, 1914), 102.

[27] Fryer, *A New Account*, 1: 270–1.

[28] Ross, *Hindu and Muhammadan Feasts*, 102–3.

[29] Abu al-Fazl, *Ain* 1: 504.

[30] Abu al-Fazl, *Akbar Nama* 2: 51.

shedding blood on their birthdays.) The next day, however, he killed three sheep with his own hand.[31] In 1632 Shahjahan led a procession to the Id-Gah where, after prayers, a goat was sacrificed,[32] and in 1644 he sacrificed the goat himself.[33] Under Aurangzeb, the emperor (or his son) sacrificed a camel, inaugurating a citywide cycle of sacrifices.[34]

Because of the many Iranian immigrants and the eclecticism of the court, Ashura was also widely celebrated in Mughal India – by Sunni Muslims as well as by Shiites. In an Imambarah (Place for the Imams) or an Ashurakhana (Ashura House) the populace gathered daily to hear stories of Husain's sufferings, view replicas of his tomb (*taziya*), and listen to elegies (*marsiya*).[35] In 1580 Father Monserrate saw mourning processions and story-tellers;[36] and in the 1620s Francisco Pelsaert reported that the populace fasted during the first nine days while on the tenth they carried replicas of the tombs of Husain and Hasan through the streets to the river.[37] In 1632 Peter Mundy witnessed mourners carrying tombs and cutting themselves with swords; no Hindus dared venture into the streets.[38] Because of a bloody riot in Burhanpur in 1669 Aurangzeb prohibited the making and carrying of tombs.[39] Despite this interruption, however, the commemoration of Ashura quickly revived. Dargah Quli Khan included brief biographies of eleven famous *marsiya-khwan* (elegy singers) in his description of eighteenth-century Shahjahanabad,[40] and in the 1820s Ashura remained the major celebration in the Shiite state of Awadh.[41]

Id-i Maulid was also routinely celebrated. In 1580 Akbar, who "... used to keep the anniversary of the Prophet's birth every year ...," entertained

[31] Jahangir, *Tuzuk* 1: 189.

[32] Peter Mundy, *The Travels of Peter Mundy in Europe and Asia*, ed. Richard Temple, 5 vols. (London: Hakluyt Society, 1907–1936), 2: 197–200.

[33] Brand, *Fathpur Sikri*, 62, 84.

[34] Niccolao Manucci, *Storia do Mogor*, trans. William Irvine, 4 vols.; reprint ed. (Calcutta: Editions Indian, 1965), 2: 325–56.

[35] Ross, *Hindu-Muhammad Feasts*, 106–9.

[36] Monserrate, *Journey*, 22.

[37] Francisco Pelsaert, *Jahangir's India: The Remonstrantie of Francisco Pelsaert*, trans. W.H. Moreland and P. Geyl (Cambridge, UK: W. H. Heffer and Sons, 1925), 73–5.

[38] Mundy, *Travels*, 2: 218–19.

[39] *Khafi Khan's History of Alamgir, being an English translation of relevation portions of Muntakhab al-Lubab*, trans. S. Moin al-Haq (Karachi: Pakistan Historical Society, 1975), 2: 213; John Fryer, who visited the Mughal Empire in the 1670s, found that celebration still outlawed. Fryer, *A New Account*, 1:273.

[40] Khan, *Muraqqa*, 66–74.

[41] Ali, *Observations on the Mussulmans of India*, 23–55, 81–91.

the high-ranking officeholders and ulama.[42] In 1633 Shahjahan invited a group of religious scholars to his palace where they recited the Quran and told stories of the prophet. Out of respect for the occasion, the emperor descended from his throne and sat on the floor among his guests. Later he distributed Rs. 20,000 to the pious and the poor.[43] In the mid-eighteenth century Dargah Quli Khan attended an Id-i Maulid celebration in Shahjahanabad at the Arab Sarai – a caravanserai housing two hundred Arab merchants. Lasting the whole night, the festivities included stories in praise of the Prophet and recitations of the Quran. The Arab merchants prepared food for the crowd, and Dargah Quli noted that the strong, heavily sweetened coffee caused nausea among many of his countrymen.[44]

Indic (Lunisolar)

Mughal India, however, was not a predominately Islamic country. Although Muslims (peasants, artisans, soldiers, officials, and officeholders) comprised perhaps 15–20 percent of the population, the vast majority of Indians were non-Muslims – mostly Hindus with a sprinkling of Buddhists, Jains, and Zoroastrians. And each of these communities had its own calendar and era, its own days of special observance. Nevertheless, given the number of converts and Akbar's strategy of "lasting reconciliation," it should come as no surprise to find that many Muslims celebrated the major Indic festivals.

Because most Indic calendars were lunisolar, the four popular festivals were seasonal: two, Vasant and Holi, occurred in early February or March, marking in India the arrival of spring; the other two, Dussehra and Diwali, came in late October or early November, signaling the end of the rainy season and the beginning of the fall harvest.

Vasant (or Basant), at the beginning of February, was dedicated to Sarasvati, the Goddess of Learning, and was celebrated with dancing and merry-making. Although both Akbar and Jahangir seem to have celebrated Vasant, the best description of the spring festival in Mughal India comes from Dargah Quli Khan. In mid-eighteenth-century Shahjahanabad there was a seven-day ceremony. On the first day the crowd (both Muslims and Hindus) gathered at the Qadam Sharif (Sacred Footprint [of the

[42] Nizam al-Din Ahmad, "*Tabaqat-i Akbari*," in H. M. Elliot and John Dowson, eds., *The History of India as Told By Its Own Historians: The Muhammadan Period*, 8 vols.; reprint ed. (Allahabad: Kitab Mahal, 1969), 5: 412.

[43] Inayat Khan, *The Shah Jahan Nama of Inayat Khan*, edited and completed by W. E. Begley and Z. A. Desai (Delhi, 1990), 118.

[44] Khan, *Muraqqa*, 47–9.

Prophet]), on the second at the tomb of Bakhtiyar Kaki (Sufi saint, d. 1263), and on the third at the mausoleum of Nizam al-din Auliya (Sufi saint, d. 1325), where they listened to music (*sama*). On the fourth and fifth days the celebrants visited the tombs of two lesser-known saints – Hazrat Rasul Numa and Hazrat Shah Turkoman. On the sixth day they were given sweets and cash at the imperial palace and at the mansions of high-ranking nobles. On the last day they danced wildly around the tomb of a certain Azizi, pouring wine over his grave.[45]

The second spring festival was Holi. Celebrated at the new moon in early March, it was a fertility ritual associated with the god Krishna (an incarnation of Vishnu). The defining activity during Holi, then as now, was the throwing of colored powder and water. A festival of reversal, where men and women mixed indiscriminately and the lower classes attacked the upper, Holi was a wild carnival. In 1580 Father Monseratte observed the combined celebration of Holi and Ashura – the lunar calendar moving the Muslim festival to the early spring. Mixed with the bloody mourning of the Muslim festival was the raucous joy of the Hindu celebration.[46] In 1623 Pietro Della Valle watched for three days as crowds of people surged through the streets, dousing one another with colored water and powder, singing and dancing. A world traveler, the Italian had witnessed an identical celebration several years earlier among the Indian merchants of Isfahan.[47] In 1665 Aurangzeb ordered the governor of Gujarat to ban Holi in Ahmadabad because of the destruction of life and property.[48]

The fall festivals – Dussehra and Diwali – celebrated the end of the monsoon rains and the coming of winter. Dussehra – late October, early November – commemorated the death of the demon king Ravana at the hands of Ram, another incarnation of the god Vishnu. Over the following ten days, according to the Ramayana (the Hindu epic), Ram and his men defeated Ravana's army and recaptured Ram's wife, Sita. Diwali, held twenty days after Dussehra, celebrated the return of Ram and Sita, after an exile of fourteen years, to their capital of Ayodhya. On Diwali houses and shops were outlined in lights and fireworks were exploded. During the twenty days between the two celebrations, the Ramayana was chanted, sung, and reenacted in villages, towns, and cities across the land. Although

45 Ibid., 43–4.

46 Monserrate, *Journey*, 21–2; Abu al-Fazl, *Ain* 2: 173; 3: 321.

47 *The Travels of Pietro Della Valle in India*, G. Havers and Edward Grey, eds. (London, 1892), 122–3; Mundy, *Travels*, 219.

48 Rahman Khan, "Awrangzib and the Hindus," *Islamic Culture* (1988): 107.

it was a Hindu festival, it was commemorated by ordinary Muslims. In the early sixteenth century, according to Sheikh Ahmad Sirhindi,

> ... during the Dewali ... the ignorant ones amongst Muslims, particularly the women, perform the ceremonies. ... They celebrate it like their own Id and send presents to their daughters and sisters.... They color their pots ... fill them with red rice and send them as presents. They attach much important and weight to this season.[49]

In Mughal India, Dussehra and Diwali signaled the beginning of the campaigning season; the rains had stopped, the crops had been harvested, and the generals were finally able to take the field. A review of the horses and elephants of the imperial and noble households marked the festivals under Akbar and Jahangir. At one point, Akbar ordered the court to discontinue mourning his mother, Maryam Makani, so that the festivities could begin.[50] Under Jahangir there were two or three days of high-stakes gambling during Diwali.[51] While Shahjahan does not seem to have celebrated these fall festivals, Aurangzeb included Diwali in his 1665 ban because of gambling and drunkenness.

Zoroastrian (Solar)

The third calendar in Mughal India was solar. On 21 March 1584 the emperor Akbar introduced the *Tarikh-i Ilahi* (Divine Era). Its calendar was the solar Zoroastrian with the familiar month and day names. Unlike the Yazdegird or Jalali months, however, Akbar gave the Divine Era months an additional adjective: Farvardin, for example, was Farvardin-i Mah-i Ilahi or the Ilahi (Divine) month of Farvardin. Also, the Divine Era calendar appended the extra five days to individual months rather than attaching them as a group to Isfand, the last month of the year.

The Divine Era festivals were described in the *Ain-i Akbari*. Most of these were Zoroastrian, and the only one to find a regular place in the Mughal calendar was Nau Ruz. Akbar wrote, "The greatest Id is Nauroz ..."[52]

Even before Akbar the Iranian New Year had been sporadically celebrated among the Mughals. In 1505 Babur wrote a poem commemorating

[49] Saiyid Athar Abbas Rizvi, *Muslim Revivalist Movements in Northern India in the Sixteenth and Seventeenth Centuries* (Agra: Agra University, 1965), 253.

[50] Muhammad Azher Ansari, "Court Ceremonies of the Great Mughals," *Islamic Culture* 35 (1961): 196; Abu al-Fazl, *Ain* 1: 226.

[51] Jahangir, *Tuzuk*, 1: 246, 268; 2: 100–1; Ansari, "Court Ceremonies," 196.

[52] *Mukatabat-i`Allami (Insha` I Abu `l Fazl) Daftar: Letters of the Emperor Akbar in English Translation*, ed. Mansura Haidar (New Delhi, 1998), 85–6.

the unusual convergence of Nau Ruz and Id-i Fitr.[53] In March 1526, however, a month before defeating the Afghans at Panipat and inaugurating Mughal rule in the subcontinent, he refused to celebrate the New Year because it was not sanctioned by the Sharia.[54] In 1544 Humayun celebrated Nau Ruz in Herat with the Safavid governor of Khurasan. His participation, however, was due more to the insistence of his Safavid protectors than to his own inclination. In 1546 in Kabul he initially banned the spring festival but, because this Nau Ruz coincided with the circumcision of the child Akbar, he eventually relented and a subdued and abbreviated celebration took place.[55]

Akbar's first Nau Ruz celebration took place in 1582. Not until the next year, however, did the ritual assume its mature nineteen-day form.[56] The first and last days were the most important: On both occasions Akbar held grand assemblies, awarding officeholders cash, horses, robes of honor, and increases in rank. During the middle seventeen days he visited the mansions of his nobles – receiving elephants, camels, Arabian horses, pearls, rubies, gold, and rich cloth.[57]

Jahangir's celebration of the Iranian New Year was modeled after his father's. A magnificent tent was erected in the Hall of Public Audience. Silk and velvet tapestries decorated the walls while carpets of silk and gold covered the floors. In the open courtyard the ranking officeholders pitched smaller tents. On the first and last days Jahangir bestowed titles, promotions, and gifts; on the other days he visited the tents of his nobles.[58] To ensure that the festivities would be lively he lifted the Islamic prohibitions on wine and drugs:

[53] Babur, *Babur Nama*, 236.

[54] Zain Khan, *Tabaqat-i Baburi*, trans. S. Hasan Askar (Delhi, 1982), xxxiii; Mohibbul Hasan, *Babur: Founder of the Mughal Empire in India* (New Delhi, 1985), 162.

[55] Ansari, "Court Ceremonies," 183–97; Sukumar Ray, *Humayun in Persia* (Calcutta, 1948): 12–13; Riazul Islam, *Indo-Persian Relations: A Study of the Political and Diplomatic Relations Between the Mughal Empire and Iran* (Tehran: Iranian Cultural Foundation, 1970), 24–36; Chopra, *Life and Letters*, 83–4; Gulbadan Begam (Princess Rose-Body), *The History of Humayun (Humayun-Nama)*, trans. Annette S. Beveridge (Delhi, 1972), 179–80.

[56] Abu al-Fazl, *Akbar Nama*, 3: 557–8.

[57] Ibid., 2: 17; 3: 589, 644; *Mukatabat-i`Allami*, 85–6.

[58] Ansari, "Court Ceremonies of the Great Mughals," 183–97; Mubarak Ali Khan, *The Court of the Great Mughals Based on Persian Sources* (Lahore, 1986), 50; Beni Prasad, *History of Jahangir* (Allahabad, 1962), 38, 89; *The Embassy of Sir Thomas Roe to the Court of the Great Mogul, 1615–19*, ed. William Foster, 2 vols. (London: Hakluyt Society, 1899), 1: 142–4.

I gave orders that whoever might wish for the intoxicating drinks and exhilarating drugs should not be debarred from using them.[59]

During his first decade Shahjahan also celebrated Nau Ruz.[60] In the mid-1630s four thousand fully dressed cavalrymen and six hundred lavishly decorated elephants lined the road to the palace-fortress; lights outlined the towers and bastions while animal fights and fireworks diverted the crowds.[61] The Peacock Throne, transferred to the imperial tent, was dubbed the "New Year's (Nau Ruz) Throne."[62] Although Shahjahan discontinued the Divine Era and its Zoroastrian calendar in his second decade, he did not abandon Nau Ruz. In fact, one of the most extravagant festivals occurred in 1650 in his new capital of Shahjahanabad.[63]

In 1659, however, Aurangzeb abolished the celebration. He wrote to his son:

In any case, as this day [Nau Ruz] is one of the festivals of the fireworshippers [*ayad-i majus*] ... there must be no repetition of this stupid act [i.e., celebrating the festival].[64]

Imperial Birthday Ceremony

As the architect of the new Mughal imperial order, Akbar, like Abbas, fashioned a new ceremonial cycle. While he celebrated the Iranian New Year, the major rituals of the Hijra year, and the seasonal festivals of the Hindus, he also created a new ceremony. The imperial birthday ceremony, drawing on the religious and cultural traditions of the subcontinent, and celebrated twice a year – on both the solar and lunar anniversaries of his birth – was an imperial ritual of integration and legitimacy.

The birthday was not a major celebration in either Islamic or Indic culture. In both traditions the major rites of passage were birth, marriage,

[59] Jahangir, *Tuzuk*, 1: 49.

[60] Ansari, "Court Ceremonies," 185–6; Khan, *Court of the Great Mughals*, 50; Khan, *Shah Jahan Nama*, xvii–xix; 37–8.

[61] *The Travels of Fray Sebastien Manrique 1629–1643. A Translation of the Itinerario de las Missions Orientals*, introduction and notes C. Eckford Luard and Father H. Hosten, 2 vols. (Oxford, 1927), 2: 190–4; Mundy, *Travels*, 2: 237–8.

[62] Chandar Bhan Brahman, "Chahar Chaman, Qawaid-i Saltanat-i Shajahan [Rules Observed During the Reign of Shahjahan]," London, British Library Persian Manuscript Collection, Or. 1892, 62–3.

[63] Muhammad Salih Kanbo Lahauri, *Amal-i Salih or Shahjahan Nama*, ed. G. Yazdani, 3 vols. (Calcutta; Asiatic Society of Bengal, 1912–1946), 3: 107–8; William Foster, ed., *The English Factories in India, 1648–50* (Oxford: Oxford University Press, 1914), 299.

[64] Dowson, *History of India*, 7: 241–2.

and death. For Muslims, however, there was one exception. By the late twelfth century the birthday of the prophet had come to be a popular public festival – Id-i Maulid or Birthday Festival. Influenced by the rapidly developing cult of the saints, the prophet's house of birth in Mecca became a place of pilgrimage, where on 12 Rabi I special ceremonies were observed.[65]

Outside the Arabian Peninsula, Id-i Maulid first appeared in the Moroccan port city of Sabta in 1236.[66] Later, Ahmad al-Mansur (1549–1603), sixth ruler of the Saadi Dynasty (1548–1641), reformulated the ceremony, changing it from a pious remembrance of the prophet's life into a glorification of his dynasty, which claimed direct descent from the prophet. Under the Allawi Dynasty (r. 1666–present), which succeeded the Saadi, Id-i Maulid became a national ritual. The new ceremony, like the Id-i Qurban (Great Sacrifice), reminded the populace of the ruler's blood line, promoting the power and legitimacy of the dynasty.[67] For the Mughal rulers, in contrast, Id-i Maulid was a purely private ceremony – recitation of the Quran, poems and stories glorifying the prophet, and gifts for the pious and learned.

Among Indian Muslims, however, a new ceremony – the *salgirah* or year knot – became popular. On the anniversary of a boy's birth, his mother tied a knot in a special piece of red or yellow string. As the years passed, the row of knots grew to match his age. Celebrated by family and friends, the festivities included music, fireworks, toys, food, and gifts. Akbar celebrated this ritual while young.[68] The knotted string aspect of the ceremony suggests two Indic rituals: the sacred thread initiation (Upanayana) of the upper-caste Hindu boy and the knotted string (*rakhi*) ceremony in which a sister ties a string of affection around her brother's wrist. Akbar also wore the *rakhi*.[69]

The third tradition that Akbar drew on was the Tula Dana (Gift of Gold). Dana (gifts for the pious or poor) referred in early Indian history to rewards given by political or military leaders to Brahmin priests at religious festivals, coronations, or military victories.[70] By the Gupta period (ca. 240–500) the Sanskrit texts began to mention the sixteen Mahadanas

[65] *Encyclopaedia of Islam*, 2d. ed s.v. "Mawlid"; Grunebaum, *Muhammad Festivals*, 67–83.

[66] *Encyclopaedia of Islam*, 2d. ed s.v. "Mawlid."

[67] Combs-Schilling, *Sacred Performances*, 10–11, 157–73.

[68] Ali, *Observations on the Mussulmans*, 215; *Qanun-i Islam*, 47. For Akbar, see Khan, *Maathir-ul-Umara* 1: 42.

[69] Ansari, "Court Ceremonies," 191–93; Abu al-Fazl, *Ain* 1: 277; Khan, *Maathir-ul-Umara*, 1: 242; Jahangir, *Tuzuk* 1: 178; Ali, *Observations on the Mussulmans*, 215; *Qanun-i Islam*, 47. For Akbar see Badaoni, *Muntakhab*, 2: 269.

[70] For a discussion, see Vijay Nath, *Dana: Gift System in Ancient India (ca 600 B.C.–300 AD)* (Delhi: Munshiram Manoharlal, 1987).

(Great Gifts).[71] While Abu al-Fazl described all sixteen in the *Ain*, it was the Tula Dana that captured Akbar's attention: a rich donor (king or governor) weighed himself against gold, distributing the proceeds to the priests and the poor.[72]

Under the Mughals the first emperor to celebrate his lunar birthday (*ruz-i viladat*) with a public weighing (*vazn*) was Humayun.[73] During his first twenty-five years (1556–1581) Akbar seems to have basically followed his father's example.[74] In 1582, however, he expanded the celebration. In that year his lunar birthday (5 Rajab) fell in the middle of the monsoon season (26 July). Heavy rains collapsed one wall of the great reservoir in Fathpur Sikri. Although the birthday festivities were interrupted and many homes and shops in the lower city were destroyed, none of the court celebrants were injured and the buildings of the imperial palace were unharmed. In thanksgiving Akbar instituted a second celebration on his solar birthday (1 Aban [17 October] of the Zoroastrian calendar).[75]

Although the lunar ceremonies of Humayun and Akbar included a weighing, the articles were not specified. In the revised ritual, however, they were. On his lunar birthday Akbar was weighed against eight, mostly ordinary, items: silver, cloth, lead, tin, fruits, sweets, vegetables, and sesame oil.[76] On his solar birthday, by contrast, he was weighed twelve times: against gold, silk, quicksilver, perfumes, copper, pewter, drugs (opium), butter, rice, milk, grains, and salt.[77]

Under Akbar the two celebrations became imperial spectacles (*jashn* or *majlis*). Celebrated in the palace-fortress or imperial camp, the music, dancing, and fireworks lasted for days. The chief household physician recorded the emperor's weight, and the items from the opposing pan were distributed to the pious and poor. The ranking officeholders received horses, cash, and increases in rank while returning gold, silver, and precious jewels.

Although Jahangir continued to celebrate both birthdays,[78] Shahjahan introduced a slight variation:

[71] Narenda Nath Bhattacharya, *A Glossary of Indian Religious Terms and Concepts* (New Delhi: Munshiram Manoharlal, 1990), 94.

[72] Abu al-Fazl, *Ain* 3: 305–7.

[73] Khwandamir, *Qanun-I Humayun (Also known as Humayun Nama)*, ed. M. Hidayat Hosain (Calcutta: Royal Asiatic Society of Bengal, 1940), 102; Ansari, "Court Ceremonies," 191; Chopra, *Life and Letters*, 86–7.

[74] Abu al-Fazl, *Akbar Nama*, 3: 539–40.

[75] Ibid., 3: 579–81.

[76] Ibid., 3: 580–1; Abu al-Fazl, *Ain* 1: 276–7.

[77] Abu al-Fazl, *Ain* 1: 276–7.

[78] Jahangir, *Tuzuk* 1: 77–8; *Embassy of Sir Thomas Roe*, 411–13.

> On the last day of Rabi I 1038 [27 November 1628] the weighing ceremony of His Majesty was celebrated in the palace at Fatehpur, on the occasion of his having attained the 38th year of his everlasting age, according to the lunar calendar. His exalted person was weighed once against gold, once against silver, and six times against other articles; which funds were then distributed among the needy. It may be observed that the tradition of the weighing ceremony (*jashn-i vazn*) was introduced by the late Emperor Akbar, who used to have his auspicious person weighed twice a year, on the occasion of his solar and lunar birthdays. At the time of the solar celebration, he was weighed twelve times: first against gold, and eleven times against other articles. On the lunar occasion, he was weighed eight times: First against silver, and seven times against other articles. The amounts which were thus realized were then distributed in charity. The late Emperor Jahangir also continued the same tradition of annual weighing ceremonies. However, the present benevolent emperor, on account of his generosity and munificence, made the innovation of weighing against gold as well as silver on the occasion of the lunar celebration, and conversely during the solar celebration as well.[79]

During the first decade of his reign, Aurangzeb celebrated his birthday according to the revised lunar ritual.[80] In 1668, however, he discontinued the weighings, and two years later he abolished the remainder of the festivities – retaining only three hours of music.[81] Nevertheless, the power of the ritual was not forgotten, even by this most ascetic and puritanical of rulers. In later years Aurangzeb allowed his sons to be weighed when they were ill or in peril – believing that the prayers of the pious would bring them health and success.[82]

The Mughal ceremonial round was complicated. Although the ritual cycles adopted by the Safavids and Ottomans diverged considerably from the practices of the early faith, in both empires the lunar liturgical ceremonies framed the year. For the Mughals, by contrast, the primacy of the canonical system was often challenged by the inherited or created ceremonies of their new order.

Unlike the Safavids, the Mughals did not radically reinterpret the basic Islamic rituals. Id-i Fitr, Ashura, and Id-i Maulid were all celebrated in the traditional manner; only Id-i Qurban was briefly altered. Where the Mughals differed from their Islamic contemporaries was in their wholesale incorporation of non-Islamic rituals. To reinforce his new imperial order

[79] Khan, *Shah Jahan Nama*, 28.

[80] Francois Bernier, *Travels in the Mogul Empire, 1656–68*, trans. Irving Brock, ed. Archibald Constable (New Delhi: S. Chand and Co., 1972), 268–71.

[81] Muhammad Saqui Mustaiid Khan, *Maasir-i Alamgir*, trans. Jadunath Sarkar (Calcutta: Asiatic Society of Bengal), 48, 59; Jadunath Sarkar, *A History of Aurangzeb*, 4 vols. (Calcutta, 1928), 3: 87.

[82] Sarkar, *Aurangzeb*, 3: 85–6.

Akbar began to publicly celebrate the four big ceremonies of the Indic year. A striking example of this syncretism was the Mughal celebration of Vasant. In mid-eighteenth-century Shahjahanabad a large crowd of Muslims and Hindus celebrated a spring festival dedicated to the Hindu goddess Sarasvati by undertaking a week-long pilgrimage around the tombs of Muslim saints.

In addition, the Mughals, unlike the Safavids or Ottomans, created a completely new ritual. The three strands of the imperial birthday ceremony reflected the composite character of the empire: Id-i Maulid was a simple ceremony from the early years of the religion, the *salgirah* was an Indian variant on Id-i Maulid, and weighing was the appropriation of an ancient Indic ritual. The ceremony symbolized Akbar's grand ambition, the wide range of goods representing the emperor's effort to incorporate the entire socioeconomic hierarchy – officeholders, soldiers, religious specialists, merchants, artisans, shopkeepers, laborers, shepherds, and peasants – into his new order.

The new ceremonial round introduced by Akbar and his successors was not the monolithic incorporation of an ancient tradition. Pieced together over time, it was fluid and dynamic – often contested and frequently revised. Akbar and Jahangir periodically declined to participate in the Id-i Qurban sacrifice while the Nau Ruz festival was altered by Shahjahan and banned by Aurangzeb. The imperial birthday celebration, followed closely by Jahangir and modified slightly by Shahjahan, was abandoned entirely by Aurangzeb.

OTTOMAN EMPIRE

The Ottoman ceremonial cycle, like the Mughal and the Safavid, was a mix of the traditional and the local. For the Ottomans, however, the rituals of the lunar Hijra calendar played a more important role. Although both the Mughals and the Safavids celebrated solar festivals from indigenous calendars, the Ottoman ceremonial innovations remained within the canonical lunar tradition. They put more emphasis on some of the minor liturgical rituals – procession to the Friday mosque and send-off of the Hajj caravan – while mostly eliminating the commemoration of Ashura – because of its ties to the hated Shiite Safavids. And the great ceremonial innovation of the dynasty – the imperial circumcision ceremony – was the magnification of a universal but indifferently celebrated Islamic rite of passage.[83]

[83] Suraiya Faroqhi, *Subjects of the Sultan: Culture and Daily Life in the Ottoman Empire* (London: I. B. Tauris, 2000), 164–84.

Of the four major ceremonies of the Hijra year, the Ottomans celebrated three. Ramadan Bayrami (Ramadan Festival) or Shukur Bayrami (Sweets Festival) marked the end of the month-long fast. In towns and cities across the empire three days (1–3 Shawwal) were set aside for fireworks, feasts, and parades. According to Ottaviano Bon, a Venetian emissary, Ahmed I (1603–1617) held an audience on 1 Shawwal in the outer court of the Topkapi Palace. Having received the great men of state, he processed to the Mosque of Hagia Sophia for the midday prayer. Afterward, the palace kitchen prepared a sumptuous banquet for the assembled guests, although the sultan himself ate alone. Later, he presented his nobles with expensive gifts (*bayramlik*) – robes, swords, and horses – and received in turn jewels, cash, and costly garments. The streets of the capital were overrun with games and merrymaking, and at night shooting flares, burning candles, and exploding missiles lit the sky above the Bosporus.[84]

Kurban Bayrami (Festival of Sacrifice) was celebrated on 10 Zu al-Hijja. In Istanbul the sultan's men sacrificed a ram in the public square and sent pieces of the bleeding animal to the great men of court. In the urban quarters the leading men of the city duplicated the sultan's sacrifice.[85]

The Prophet Muhammad's birthday (12 Rabi 1) had been celebrated at the Ottoman court from the early fifteenth century. The most popular *mawlid* (panegyric poem) in Ottoman Turkish was composed by Suleiman Celebi (d. 1421), and in mid-sixteenth-century Istanbul the celebration at the Suleimaniye Mosque included a prayer service, the chanting of the Quran, and the recitation of a *mawlid* (*mawlid-nami-i Nebil*). Food and drink were served to the assembled crowd.[86] Because the Hajj caravan usually returned to Istanbul about the first of Rabi 1, some two months after the end of the sacred month, a letter of greeting from the Sharif of Mecca was often read during the birthday festivities.[87]

In keeping with his self-image as caliph and protector of the holy cities, the Ottoman sultan organized an elaborate ceremony to celebrate the departure of the Hajj caravan. Unlike the other liturgical festivals, however, its timing varied: Cairo was 798 miles (1,285 km) from Mecca, Damascus 861 miles (1,385 kms), and Istanbul 1,489 miles (2397 km). The Cairo ceremony had an illustrious heritage, the Ottoman celebration a continuation of the earlier Mamluk tradition. On taking his leave from the governor

84 Ottaviano Bon, *The Sultan's Seraglio: An Intimate Portrait of Life at the Ottoman Court*, introduced and annotated by Godfrey Goodwin (London: Saqi Books, 1996), 113–16.
85 Ibid., 136; Lewis, *Everyday Life*, 133.
86 *Encyclopaedia of Islam*, 2d. ed., s.v. "Mawlid."
87 Lewis, *Everyday Life*, 121.

of Egypt, the caravan commander placed a large camel carrying a *mahmil* (an empty palanquin symbolizing the absent sultan) at the head of the procession.[88] The Istanbul caravan, because of the greater distance, had to depart much earlier. Camels and carts were loaded with provisions for the pilgrims and gifts for the Sharifs of Mecca and the poor of the city. Two large camels, covered with rich cloth, led the convoy. The first carried an intricately wrought *mahmil*, and the second bore an empty silver saddle covered in green velvet, symbolic of the absent prophet. A large contingent of soldiers protected the travelers.[89]

The Ottoman addition to the ceremonial cycle, unlike the Mughal or the Safavid, was quintessentially Islamic. The imperial circumcision ceremony celebrated a universal practice, male circumcision being the primary outward sign of membership in the faith. Although common in first millennium Arabia, circumcision was not mentioned at all in the Quran.[90] The hadith (sayings and stories of the prophet), however, are another matter. One has Muhammad coming from the womb already circumcised, and, according to another, the prophet said, "Let him who becomes a Muslim be circumcised, even if he is old."[91] While nearly universal in the early community, the practice was mostly ignored by the jurists. Al-Bukhari (810–870) did not mention circumcision at all in the nine thousand pages of his treatise; al-Ghazali (1058–1111) gave the practice seven lines in two thousand pages; and the seventeenth-century *Fatawa-i Hindiyya* allowed the topic only one-third of a page out of three thousand.[92] Even when it was discussed, there was no agreement: Al-Ghazali decided that it was a coming-of-age rite and should be done when pubic hair first appeared, the *Fatawa-i Hindiyya* thought seven to twelve the appropriate age, while in Madagascar the operation was done on babies of one or two.[93]

In most of the Islamic world the young boy was feted by family and friends, receiving clothes, sweets, and other gifts. For the Ottoman ruling family, however, a private event was not enough. The first elaborate public ceremony seems to have been organized by Murad I (1359–1389) for his

[88] Suraiya Faroqhi, *Pilgrims and Sultans: The Hajj under the Ottomans, 1517–1683* (London and New York: I. B. Tauris and Co., 1994), 37–9.

[89] Ibid.; Lewis, *Everyday Life*, 122–3.

[90] *Encyclopaedia of Islam*, 2d. ed. s.v. "Khitan"; David Gollaher, *Circumcision: A History of the World's Most Controversial Surgery* (New York: Basic Books, 2000).

[91] Encyclopaedia of Islam, 2d. ed. s.v. "Khitan"; Abdelwahab Bouhdiba, *Sexuality in Islam* (London: Routledge, 2007), 175; Jonathan Berkey, "Circumcision Circumscribed: Female Excision and Cultural Accommodation in the Medieval Near East," *IJMES* 28 (1996): 19–38.

[92] Bouhdiba, *Sexuality in Islam*, 175.

[93] Ibid., 87–8.

three sons – Beyazid, Yakub, and Savci.[94] In 1457 Mehmed II (1451–1481) arranged a two- to three-week festival in Edirne for two sons (Beyazid II [his successor] and Mustafa),[95] and in 1472 an even longer fifty-day event commemorated the occasion for Cem.[96] In 1491 Beyazid II (1481–1512) put together a double celebration – the circumcision of his grandson and the marriage of his daughter.[97] Suleiman the Magnificent (1520–1566) hosted two lavish affairs: in 1530 a three-week celebration marked the circumcision of Mehmed, Suleiman, and Mustafa, and in 1539 a two and a half week affair commemorated the event for Beyazid and Jahangir.[98]

The greatest imperial festival (*sur-i humayun*) in the history of the dynasty, however, was organized in 1582 by Murad III (1574–1595) to celebrate the circumcision of his sixteen-year-old son Mehmed – the future Mehmed III (1595–1603).[99] This celebration was significant not only for its length (over fifty days) but also for its three festival books (*surnames*): Lokman's description in his history of Murat III (*Shahinshahname I*) included forty-two miniatures;[100] Mustafa Ali's *Jami al-Buhur der Mecalis-i Sur* (*Gathering of the Seas on the Scenes of the Festival*) was not illustrated; while Intizami's *The Festival Book of Murat III*, featuring 430 miniatures by the court painter Osman, was the most famous.[101]

Staged in the At Meydan, the event was opened by Murad III on 1 June 1582. He led a great procession of Janissaries, ranking officials, and military officers from the Topkapi Palace to the piazza.[102] The remaining fifty days,

[94] Metin And, *A History of Theater and Popular Entertainment in Turkey* (Ankara: Forum, 1964); Esin Aka Atil, "Surname-I Vehbi: An Eighteenth Century Ottoman Book of Festivals," (Ph.D. Diss., University of Michigan, 1969), 387–92.

[95] Franz Babinger, *Mehmed the Conqueror and His Time* (Princeton, NJ: Princeton University Press), 148–9.

[96] Atil, "Surname-I Vehbi," 387.

[97] Ibid., And, *History of Theater*, 17.

[98] Atil, "Surname-I Vehbi," 387–8; Fanny Davis, *The Palace of Topkapi in Istanbul* (New York: Scribner, 1970), 9–10.

[99] For an introduction, see Robert E. Stout, "The Sur-I Humayun of Murad III: A Study of Ottoman Pageantry and Entertainment," (Ph.D. diss., Ohio State University, 1966); Derin Terzioglu, "The Imperial Circumcision Festival of 1582: An Interpretation," *Muqarnas* 12 (1995): 84–100; s.v. *Encyclopaedia of Islam*, 2d. ed. s.v. Murad III; Christine Woodhead, "Murat III and the Historians: Representations of Ottoman Imperial Authority in Late Sixteenth Century Historiography, "in Karateke, ed., *Legitimizing the Order*, 85–98.

[100] Atil, "Surname-I Vehbi," 390.

[101] Stout, "The Sur-I Humayun," 9–18; Terzioglu argues that the traditional attribution of the text to the historian Lokman b. Huseyn al-Ashuri is wrong, and that the author was a clerk in the imperial divan. "Imperial Circumcision Festival," 84.

[102] See map in Stout, "The Sur-I Humayun," 58, 73–105; Terzi, "Imperial Circumcision Festival," 86.

until the actual circumcision, were filled with guild processions (more than 150),[103] music (drums, trumpets, fifes, pipes, tambourines, harps, and flutes), dances (religious, erotic, and comic), circus acts (rope-walkers, fire-eaters, conjurers, strong men, clowns, jugglers, mimes, puppets, and tumblers), mock battles (St. George and the Dragon and the recent Ottoman victory over the Safavids), and fireworks.[104] Murad III, however, not Prince Mehmed was the ceremonial center. Ambassadors, imperial officers, religious dignitaries, and guild leaders offered gifts and congratulations, while the secluded sultan returned coins or food.[105] Ordinary spectators received rice, bread, and mutton from the imperial kitchen.[106]

Mehmed's circumcision was not the only one celebrated during the fifty-day period. In the At Meydan, 150 Christian converts a day were circumcised while across the empire thousands of boys underwent the operation, their festivities merging with the great event in the capital.[107]

While the 1582 festival was the defining example of its type, Murad III was not alone in organizing an elaborate ceremony. In 1675 in Edirne Mehmed IV (1648–1687) celebrated the circumcision of two sons – the future Ahmed III (1703–1730) and the future Mustafa II (1695–1703)[108] – and in 1696 Ismail Pasha, governor of Egypt (1695–1697) organized a massive event for his son.[109] The last of the elaborate festivals was the fifteen-day event organized in 1720 by Ahmed III (1703–1730) for his four sons: Suleiman, Mehmed, Mustafa, and Beyazid.[110] For his festival book Ahmed organized a competition: the winners were Vehbi, a poet, and Rasid, the court chronicler, but the fame of *Surname-i Vehbi* was due to the

[103] Stephane Yerasimos, "The Imperial Procession: Recreating a World's Order," 2, 7. http://www.geocitites.com/surnamei_vehbi/yerasimos.html9/30/2008.

[104] Stout, "The Sur-I Humayun," 106–25; Terzi, "Imperial Circumcision Festival," 86.

[105] Stout, "The Sur-I Humayun," 67–8; Yerasimos, "The Imperial Procession," 5.

[106] Stout, "The Sur-I Humayun," 73–105; Yerasimos, "The Imperial Procession," 6.

[107] Stout, "The Sur-I Humayun," 269–76; Terzioglu, "The Imperial Circumcision, 84; Gisela Prochazka-Eisl, "Guild Parades in Ottoman Literature: The Surname of 1582," in Suriaya Faroqhi and Randi Deguilhem, eds. *Crafts and Craftsman of the Middle East: Fashioning the Individual in the Muslim Mediterranean* (New York: IB Tauris, 2005), 41–54.

[108] Faroqhi, *Subjects of the Sultan*, 165; Yerasimos, "Imperial Procession," 2–3, 7–8.

[109] Doris Behrens-Abouseif, *Egypt's Adjustment to Ottoman Rule: Institutions, Waqf, and Architecture in Cairo, Sixteenth and Seventeenth Centuries* (Leiden: Brill Academic Publishers, 1994), 65–6.

[110] For a discussion see Atil, "Surname-I Vehbi"; Esin Atil, " The Story of an Eighteenth-Century Ottoman Festival," *Muqarnas* 10 (1993): 181–200; Yerasimos, "The Imperial Procession;" Faroqhi, *Subjects of the Sultan*, 165–8; Fanny Davis, *Palace of Topkapi*, 158–59.

137 miniatures contributed by the famous court painter Abdulcelili Chelebi "Levni."[111]

Why did the Ottomans redefine the Islamic custom of male circumcision, transforming a mostly domestic ritual into a large public festival? What role did these elaborate ceremonies play in the political culture of the dynasty?

In Safavid Iran and Mughal India quite different traditions prevailed. For the Safavids circumcision, for rulers and commoners alike, was a predominately family affair. Accompanied by feasts and merry-making, the operation was scheduled (at no particular age) by the parents and performed by the neighborhood barber. At this time the young man received his formal name.[112] Unlike the Ottomans, the Safavid rulers did not publicly commemorate the circumcisions of their sons. Iskandar Munshi, for example, in his three-volume history of Shah Abbas I, devoted a section to the children of Shah Tahmasp, Abbas's grandfather. Nine sons, eight daughters, three nephews, and five grandsons (sons of Muhammad Khudabanda, Tahmasp's successor and Abbas's father) were briefly described. At no point – including the five pages devoted to the young Abbas – was circumcision mentioned.[113]

In India, as in Iran, the circumcision of Muslim males was not a coming-of-age rite. The parents chose a time that was convenient – to coincide with the marriage of a sister or the circumcision of a brother – and thus the age at which the operation was performed varied from three to fourteen.[114] The Mughal emperors, however, in contrast to the Safavids, did organize celebrations. The longest was the seventeen-day festival in 1545 for the three-year-old Akbar. Humayun, Akbar's father, had been defeated by the Afghans in 1540 and had been forced into exile in Afghanistan and Iran. After capturing the city of Kabul in the late fall, he scheduled the circumcision of his infant heir for the early spring. As with the Ottomans, the celebration centered on the imperial father not the princely son. At the daily audiences Humayun received gifts and congratulations from office-holders and ambassadors (including one from Shah Tahmasp) while

111 Atil, "Surname-I Vehbi," 7, 54; Yerasimos, "The Imperial Procession," 5–6; Nurhan Atasoy and Filiz Cagman, *Turkish Miniature Painting* (Istanbul, 1974), 73–9.

112 Chardin, *Voyages*, 9: 191–3; Fryer, *A New Account*, 3: 138; Du Mans, *Estate de la Perse*, 77.

113 Iskandar Munshi, *Tarikh-e Alamara-ye Abbasi: History of Shah Abbas the Great by Eskandar Beg Munshi*, trans. Roger M. Savory, 2 vols. (Boulder, CO: Westview Press, 1978),1: 206–21.

114 Ross, *Hindu and Muhammadean Feasts*, 104; Chopra, *Life and Letters*, 89–95.

dispensing robes, promotions, and grants of land. The description of Akbar's ceremony in 1545, unlike the *Surname-i Humayun* of 1582, had a primarily domestic focus. Instead of a breathless account of parades and entertainments, Abu al-Fazl's narrative revolved around the three-year-old infant's immediate recognition of his mother, Hamida Banu Begam, whom he had not seen for over two years.[115]

In 1573, twenty-eight years after his own circumcision and seventeen years after ascending the throne, Akbar arranged a celebration in Fathpur Sikri for his three sons: Salim (later Jahangir) was four, Murad three, and Daniyal one.[116] Lasting several days, the festivities were marked by feasts, the exchange of gifts, the weighing of the emperor, and precious articles distributed to the poor, the learned, and the pious.[117] Eighteen years later in 1591, having completed his reorganization of the Mughal state and having adopted a tolerant though occasionally critical stance toward the social and religious beliefs and practices of his subjects, Akbar returned to circumcision. In his instructions to the city magistrate (*kotwal*) the emperor ordered: "He should not suffer a woman to be burnt against her inclination [i.e., the custom of sati] nor a criminal deserving of death to be impaled, nor anyone to be circumcised under the age of twelve."[118] And in "The Sayings of His Majesty," Abu al-Fazl recorded: "It is a remarkable thing that one should insist on the ceremony of circumcision for children who are otherwise excused from the burden of all religious obligations."[119] Akbar appears not so much anti-Muslim (as the conservative Badauni charged) as skeptical. As a marker of religious identity, shouldn't the operation (serious and painful) be postponed to the age of reason?

Akbar's order and opinion notwithstanding, there is no evidence of any drastic change among the Muslims of early modern India. His son Jahangir, circumcised at the age of four, scheduled the operation for Shahjahan, his son and successor, at the same time.[120] And Muhammad Akbar, Aurangzeb's son, was four when he was circumcised. For the Mughal heir apparent, and perhaps for other princes as well, the ceremony marked an important transition. Circumcision was the first step in the *maktab*

[115] Abu al-Fazl, *Akbar Nama*, 1: 483–9.
[116] Ibid., 1: 103.
[117] Ibid.; Srivastava, *Akbar*, 1: 156.
[118] Abu al-Fazl, *Ain* 2: 45. See also Badaoni, *Muntakhab*, 2: 388.
[119] Abu al-Fazl, *Ain* 3: 441.
[120] Banarsi Prasad Saksena, *History of Shahjahan of Dihli* (Allahabad: Central Book Depot, 1968), 2.

(school) rite – when the young prince left the harem and began his formal schooling.[121]

In the literature the explanation for the creation of a new Ottoman imperial ritual was the ending of formal sultanic weddings. From the late thirteenth to the mid-fifteenth century, Ottoman rulers contracted legal marriages with nearby European and Anatolian rulers. With the death of Mehmed the Conqueror (1451–1481), however, the Ottomans, with their claim to world dominion, judged no other dynasty worthy of a marital alliance. Thereafter, the imperial harem contained only slave concubines.[122] At this point, it is argued, the imperial family redefined the circumcision celebration. Elaborated and expanded, it was elevated to the status of a dynastic ritual, filling the void created by the elimination of sultanic weddings.[123]

When looked at more closely, however, the substitution theory seems less persuasive. In the first place, there were sizeable circumcision celebrations before the late fifteenth century. Murad I had a public ceremony for his three sons in 1387 and Mehmed II had lavish celebrations in 1457 (two sons) and 1472 (one son). But, more important, a proper interpretation needs to examine the ritual in a broader context, as part of a ceremonial reordering during a period of dynastic change. Under Mehmed II the sultan began to retreat from public view. According to his *Kanunname*, the ruler was to seclude himself in the Topkapi Palace, appearing in public only twice a year – at Ramadan Bayrami and Kurban Bayrami. The aim was to limit the sultan's public appearances in order to magnify their effect. A strict protocol came to govern all imperial ceremonies.[124]

With the absence of the emperor from public ceremonies, the Ottomans redefined and elevated two life-cycle rites. The marriages of princesses and the circumcisions of princes became more public and more lavish. In the process the ceremonial face of the dynasty changed, attention shifting from the sultan to his sons and daughters. Marriages and circumcisions became, in a way unique to the Ottomans, the definitive coming-of-age rituals, the principal rites of passage from adolescence to adulthood. Promoted to a new place in the ceremonial round, they reminded the populace of the power and authority of the increasingly absent sultan.

[121] Ibid.

[122] Leslie Pierce, *The Imperial Harem: Women and Sovereignty in the Ottoman Empire* (New York: Oxford University Press, 1993), 28–30.

[123] Faroqhi, *Subjects of the Sultan*, 164–5.

[124] Woodhead, "Murad III," 88; *Encyclopaedia of Islam*, 2d. ed s.v. "Kanunname."

In their new role the two rituals were complementary. The Ottoman word "*sur*," meaning a feast or celebration or, more specifically, a circumcision or wedding feast, came from the Persian "*sur*," a wedding. In the early modern Ottoman empire the circumcision ceremony of the prince was the equivalent of the marriage celebration of the princess: For the boy it was "*sur-i hatan* (male celebration)" and for the girl "*sur-i cihaz* (female celebration)."[125] Both commemorated a liminal event, crossing the threshold separating adolescence from adulthood, signifying that both imperial offspring were ready to assume their proper and rightful places in the wider world.[126]

For Muslim girls in the Ottoman, as in the Mughal and Safavid empires, it was marriage not circumcision that was the defining coming-of-age ritual. While female excision was practiced in certain parts of the Islamic world (chiefly Egypt and North Africa), there is very little evidence of the custom in the three early modern Islamic empires.[127] After the reign of Mehmed II, the marriages of imperial daughters, sisters, and nieces became increasingly important and their "*surs*" longer and more elaborate. The young women were given to high-ranking slave members of the imperial household, and the *damad* (son-in-law) came to assume greater responsibility in state administration. In 1524 Suleiman celebrated the marriage of his sister to Ibrahim Pasha in an eight-day festival, in 1586 Murad III married his daughter Ayse to Kanijeli Ibrahim Pasha, and in 1593 he married Fatma to Halil Pasha in a seven-day ceremony. Ahmed I (1603–1617) married three of his daughters to men who were eventually promoted to grand wazir, and Ahmed III in 1709 organized a nine-day celebration for the marriage of his daughter Fatma to Silahdar Ali Pasha.[128] For the Ottomans the virtue of these marriages lay in the bonds of loyalty created between the sultan and his high-ranking and ambitious subordinates.

Of the two surs, however, it was the "*sur-i hatan* (male celebration)" that came to predominate. Within the broad outlines of Islamic practice, the Ottomans created a new variant, marking a transition to a different stage of life. Although the Jewish rite (on the eighth day) marked the transition

[125] A.D. Alderson, *The Structure of the Ottoman Dynasty* (Oxford: Clarendon Press, 1956), 104.

[126] Stout, "The Sur-I Humayun, 27–8; Alderson, *The Structure of the Ottoman Dynasty*, 104; Bouhdiba, *Sexuality in Islam*, 134.

[127] Berkey, "Circumcision Circumscribed."

[128] Atil, "Surname-I Vehbi," 387–91; Pierce, *Imperial Harem*, chs. 2–3; Stephen P. Blake, "Returning the Household to the Tributary Empire Model: Gender, Succession, and Ritual in the Mughal, Safavid, and Ottoman Empires," in Peter Fiber Bang and C.A. Bayly, eds., *Tributary Empires in Global History* (London: Palgrave Publishing, 2011), ch. 13.

from womb to family and faith, the Mughal imperial practice (at age four) signaled the movement from harem to school. The Ottoman imperial custom, by contrast, commemorated the passage from adolescence to full maturity. The young prince, often the heir apparent, was initiated into the adult world, ready to assume the roles of husband, father, believer, and ruler.

Because the year chosen for any particular ceremony depended on a variety of factors – other princes to circumcise, princesses to marry, or military defeats to hide (the Ottoman failure at Vienna in 1529, for example)[129] – the age of the primary celebrant could vary. Nevertheless, in contrast to Mughal and Safavid practice, the Ottoman rite was not only public but unmistakably adolescent. In 1457 Beyazid II was ten, in 1472 Cem was thirteen, in 1530 Mustafa was sixteen, in 1539 Beyazid was fourteen, in 1582 Mehmed was sixteen, in 1675 Mustafa II was twelve, and in 1720, to illustrate the difficulties of coordination, Ahmed circumcised two future sultans – Mahmud I (1730–54) was twenty-four and Mustafa II (1757–1774) was three. Ottaviano Bon put the usual age of the prince at fourteen,[130] and Murad III's invitation to the 1582 ceremony contained an unmistakable allusion to sexual maturity: "... the plant of his existence ... had some happy enlargements in the garden of virility and force ... the vegetative knob which is the chief of the reproductive faculties ..."[131]

In reorganizing the circumcision ceremonies of their sons, the Ottoman sultans of the early modern era transformed a private domestic ritual into a public dynastic event. The new festival was intended to strengthen their authority and legitimacy. Although secluded, the ruler was far from uninvolved in the new ritual. He chose the date and directed the preparations. In the parades, audiences, and entertainments he was the focus. Through subordinates he accepted the offerings of wazirs, generals, religious officials, guild leaders, and entertainers, on the one hand, while returning coins, robes, and food, on the other. These symbolic exchanges reconfirmed the hierarchical order of Ottoman society. Participation in the audiences, processions, and entertainments was an act of loyalty, a reaffirmation of the subject's allegiance to his ruler.

In each of the three early modern Islamic empires the ruling dynasties created new ceremonial cycles. While these new ritual arrangements did not ignore the traditional festivals of the Hijra year, they did – in their

129 Pierce, *Imperial Harem*, 170

130 Bon, *Sultan's Seraglio*, 91.

131 Stout, "The Sur-I Humayun," 45–7.

additions and subtractions – reflect the unique form that Islam had taken in each state. Because of the ethnic and cultural diversity of early modern India, the Mughal ceremonial order was the most complex. Due to the strong Iranian representation among the Mughal elite, Ashura was the most popular of the four Hijra ceremonies. In addition, the heavy Persian cultural and artistic influence at the Mughal court strongly affected Akbar's choice of a calendar for his new solar era. It was Zoroastrian – the Divine Era year began at the Vernal Equinox, the day and month names were nearly identical to the ancient Iranian, and the five extra days were appended, although with slightly different names and locations. All of the traditional Zoroastrian festivals appeared on Akbar's new calendar but only Nau Ruz was regularly celebrated. The emperor even enlarged the traditional ceremony – extending it from eight days (under the Safavids) to nineteen. Akbar also introduced a completely new ritual. The imperial birthday ceremony was an amalgam: part Islamic, part Indian Muslim, and part Indic.

The Safavid ceremonial round, on the other hand, was not as eclectic. A dynasty in a country that had been conquered by the early Umayyads, the Safavids fashioned a ceremonial order that reflected both their ancient Zoroastrian heritage, on the one hand, and their recent adoption of Shiism, on the other. Although Nau Ruz remained extremely popular – marking the beginning of spring and the first day of the solar year – it was gradually replaced as the centerpiece of the ritual year by Ashura. The dynasty's promotion of the ten-day, emotionally charged commemoration influenced the two other major ceremonies – adding a public distribution of the bloody camel to Id-i Qurban while interjecting a glorification of Ali into Nau Ruz.

Although the Ottomans (unlike the Mughals or Safavids) created a ceremonial cycle that was entirely Islamic, they did not limit themselves to the four festivals of the Hijra calendar. Because of their long-standing rivalry with the Safavids, they eliminated Ashura from their ritual repertoire. They added, however, two other canonical festivals that were either underplayed or ignored in the other two states: the imperial circumcision ceremony and the ceremony celebrating the departure of the Hajj caravan.

In all three empires the creation of a new ceremonial cycle was part of a larger project – the construction of an imperial idiom that would undergird the authority and legitimacy of the new dynasty. For the Mughals, the task was complicated by the ethnic and religious complexity of early modern India. Akbar's ritual had to integrate and coordinate the festivals of Muslims and non-Muslims alike. For the Safavids, the aim of ceremonial reorganization was to promote a new dynasty with a new source of legitimacy. Abbas's goal was to convert the Iranian populace to Shiite Islam while at the same time retaining their ancient belief in the semidivine nature of kingship. For the Ottomans, without the Timurid ancestry of the Mughals or the spiritual aura of the Safavids, elevating their dynasty in the hierarchy of Sunni Islam was the chief path to enhanced authority and legitimacy – thus the ceremonial importance given to the departure of the Hajj caravan and the circumcision of imperial princes.

Finally, in early modern Islam the ritual cycle was not the continuation of an ancient tradition. Rather, in all three empires the ceremonial order was dynamic – ceremonies were created, contested, revised, and occasionally abandoned.

4

Chronology: Era

From the chronological perspective the study of time is the study of the year. For the three early modern empires Islamic chronology presented two problems: First, how to order and administer an agrarian state using a lunar era? Second, how to manage the millenarian hysteria thrown up by the chronological milestones (1583 and 1591) of the late sixteenth century?

In the first decade after the prophet's death, the early Muslim community established a new era. In his book on chronology al-Biruni wrote:

> Era means a definite space of time, reckoned from the beginning of some past year, in which either a prophet, with signs and with a proof of his divine mission, was sent, or a great and powerful king arose, or in which a nation perished by a universal destructive deluge, or by a violent earthquake and sinking of the earth, or a sweeping pestilence, or by intense drought, or in which a change of dynasty or religion took place, or any grand event ... which does not happen save at long intervals and at times far distant from each other. By such events the fixed moments of time [the epochs] are recognized. Now, such an era cannot be dispensed with in all secular and religious affairs.[1]

The epoch of the Hijra Era was 622, and its years were lunar – 355 days, twelve months of twenty-nine or thirty days each.

After the death of Ali (656–661), the prophet's son-in-law and the last of the four rightly guided caliphs (or deputies), a fledgling Muslim state emerged. The Umayyad Dynasty (661–750), headquartered in Damascus, governed a rapidly expanding territory that extended from North India in

[1] al-Biruni, *Chronology*, 16.

the east to North Africa and the Iberian Peninsula in the west. As the spoils of conquest drew to an end, the need for a regular system of taxation became increasingly urgent. In an agrarian economy, the lunar Hijra Era was seriously deficient. Taxes had to be collected soon after the annual harvest, and with a lunar era the date of collection did not fall at the same point in the seasonal cycle. Any date in the Hijra Era year – for example, 1 Muharram (New Year's Day) or 10 Ramadan – regressed at the rate of about eleven days against the seasons. Thus, Abd al-Malik (685–705), the third Umayyad ruler, decided, for administrative and fiscal reasons, to adopt a new solar era. However, this new era could not be allowed to challenge the primacy of the Hijra. Thus, the Kharaji or Taxation Era (AK), with the 365-day Zoroastrian calendar (twelve months of thirty days plus five at the end) had to be designed to track the liturgical era. In the early years of its adoption this meant that the solar era numbering sequence had to be adjusted so that it would overlap the Hijra, 68 AK matching 68 AH, and so on. But, because of the eleven-day difference, after thirty-two years the two eras would diverge. Left uncorrected, this discrepancy would have introduced confusion into official documents and records. It would also have left the peasant at the mercy of unscrupulous tax collectors. After paying the taxes for the current solar year, the cultivator could be dunned some months later for another payment – due because the lunar year had changed. To avoid this confusion the Kharaji Era had to be periodically recalibrated, the numbering sequence advanced. This process was called *izdilaq* (sliding) in Arabic and *tahvil* (changing) in Persian. Thus, in every Kharaji Era century the years after the thirty-second, the sixty-fourth, and the ninety-sixth were eliminated – 34 AK following 32, 66 following 64, and 98 following 96.[2]

As time passed, however, this method of remedying the deficiencies of the liturgical era proved increasingly unsatisfactory. It was difficult for administrators, tax collectors, record keepers, and historians to remember the "sliding" required to keep the Kharaji Era synchronized with the Hijra. Thus, by the early modern period the Mughal and Safavid empires had abandoned the Kharaji Era and had adopted one or more solar eras, with numbering sequences separate from the Hijra. The Ottomans, on the other hand, chose not to employ one of the several independent solar eras, deciding instead to continue with a taxation era that required periodic

[2] The first documented example was in 242 AH/856–57 CE during the reign of the Abbasid Caliph al-Mutawakkil (232–47 AH/846–62 CE). Taqizadeh, "Various Eras and Calendars," 903–16; Encyclopaedia *of Islam*, 2d. ed., s.v. "Tarikh," 263.

adjustment. In all three empires, however, the new chronologies created problems. How to integrate the new eras into the existing systems of tax collection and revenue disbursement? How to date historical accounts and to accurately translate dates from one era to another?

SAFAVID EMPIRE

Fiscal

In pre-Islamic Iran under both the Achaemenid (559–330 BCE) and Sassanid (224–651 CE) dynasties the solar Zoroastrian calendar named and grouped the days. Both empires, however, employed a regnal system for numbering the years, the crowning of a new ruler inaugurating a new era. Thus, a battle or a birth would be in "the twelfth year of Darius" or "the seventh year of Cyrus." This scheme had two advantages: it was understood throughout the entire empire and could extend over several rulers. The disadvantage, however, was the random beginning of each era – on whatever day the new king happened to ascend the throne. The Achaemenids eventually solved this problem by designating New Year's Day of the accession year as the epoch of the new era. This subordination of the regnal year to the civil, followed by the Sassanids as well, was done in order to make record keeping, tax collection, and historical accounts predictable and accurate.[3]

When, however, the invading forces of Umar (634–644), the second caliph, defeated the Sassanid army at the battle of Nihavand in 641, the regnal system of historical accounting came to an end. Yazdegird III (632–651), the last Sassanid ruler, had come to the throne in 632 and so the era that he inaugurated did not end with his death.[4] The Yazdegird Era (epoch 632), found in the *Zij-i Malik Shahi*, the *Zij-i Ilkhani*, and the *Zij-i Sultani or Gurgani*, employed the 365-day Zoroastrian calendar. Since, however, there was no provision for the extra quarter-day, the date of the New Year slowly regressed. In the late Sassanid period the New Year began at the Summer Solstice (21 June) but during the first centuries of Islamic rule it was set at the Vernal Equinox (21 March).[5] By the middle of the eleventh century, therefore, Nau Ruz had retreated to late February. As

[3] Ehsan Yarshater, ed., *The Cambridge History of Iran 3/2: The Seleucid, Parthian and Sasanian Periods* (Cambridge: Cambridge University Press, 1983), 792–815.

[4] *Encyclopaedia of Islam*, 2d. ed., s.v. "Tarikh."

[5] Taqizadeh, "Various Eras," 917.

a result, the Seljuq ruler Malik Shah ordered his astronomers and astrologers to produce a new solar era, one that would both correct the quarterday error and return Nau Ruz to the Vernal Equinox. The epoch of the Jalali Era was 21 March 1079 AD, and its 365-day Zoroastrian calendar included a leap year – every fourth year six days rather than five were added to the basic 360. This new era kept the seasons and the calendar synchronized, and by the Safavid period it was called the Solar Jalali (Khurshid-i Jalali) Era.[6]

The final era regularly employed by the Safavid astrologers, historians, and administrators was the Turkish Twelve-Year Animal Era. Also called the Chinese-Uighur Animal Era, it was brought to Iran by the Mongols. Under Genghis Khan (ca. 1162–1227) the Mongols came to loosely rule an empire that covered most of Eurasia – from China in the east to Russia and Eastern Europe in the west. The Mongols' sociocultural system had a magpie look, adopting elements from the many states they had conquered. From the Chinese they took the twelve-year lunisolar era – twelve months of twenty-nine or thirty days with an extra month intercalated every three years. The version adopted by the Mongols came via the Uighurs, a Turkish-speaking central Asian people who served as the cultural conduit between the Chinese and the Mongols. The year began circa 27 January (15 degrees Aquarius) and the month and year names were Turkish.[7] Under the Safavids, however, the era became solar: the year began at the Vernal Equinox and the names of the months were Zoroastrian. The twelve 365-day years, however, retained the names and order of the twelve animal signs of the Chinese zodiac:

1. Sichuan (Mouse);
2. Ud (Cow);
3. Ba-Pars (Leopard);
4. Tawishqan (Hare);
5. Lui (Crocodile);
6. Ilan (Snake);
7. Yunt (Horse);
8. Qui (Sheep);
9. Bichi (Monkey);
10. Takhaqui (Fowl);

[6] Ibid., 108–17.

[7] Charles Melville, "The Chinese Uighur Animal Calendar in Persian Historiography of the Mongol Period," *Iran* 32 (1994): 83–98; Louis Bazin, *Les Systemes chronologiques dan le monde turc ancient* (Budapest and Paris: Bibliotheca Orientales Hungaria, 1991).

11. It (Dog);
12. Tanguz (Hog).[8]

Since the era recycled every twelve years it had to be linked in any system of historical chronology to another longer-lasting era. Interestingly enough, the Turkish Animal Era was similar to a medieval European system of chronology. From about 312 the Europeans grouped their years into fifteen-year indiction cycles, each year identified by a number (eight or twelve or fifteen), whereas the eras themselves, like the Turkish Animal, were not numbered.[9]

For the Safavids the fiscal problem was easily solved. Because of their experience with solar chronologies (the Yazdegird, Jalali, and Turkish eras), crafting an accounting system that would accurately track agrarian revenues and disbursements was not difficult. To identify the year the administrators and scribes paired the animal year of the Turkish Era with a Hijra Era date.[10] If further accuracy was required, the Zoroastrian days and months of the Jalali Era were added. For example, a document from the reign of Shah Tahmasp established the rules for distributing water from the Zayanda River: access to one canal was divided into fifteen-day segments for the 3.5 month period from 1 Tir to 18 Mehr.[11]

Dating

For historians of Safavid Iran chronology was a central concern. Given the variety of eras available, how were chroniclers to organize their accounts? What dating system or systems should they employ? What criteria should govern their choices? To begin to answer these questions let us look at four historians from the reign of Shah Abbas I: Mahmud b. Hidayat Allah, Afushtah-i Natanzi; Jalal al-Din Muhammad Yazdi; Iskandar Beg Munshi; and Mirza Beg b. Hasan Junabadi.

Although the history of Iskandar Beg (ca. 1560–1632) was not the first, it has become, because of its English translation, the best known. Also, as it

[8] *Encyclopaedia Iranica*, s.v. "Calendar," 371.

[9] Deborah Mauskopf Deliyannis, "Year Dates in the Early Middle Ages," in Chris Humphrey and W. M. Ormrod, eds., *Time in the Medieval World* (Suffolk: York Medieval Press, 2001), 5–22.

[10] Bert Fragner, "Social and Economic Affairs," in Jackson ed., *The Cambridge History of Iran*, 6: 556.

[11] Ann K. S. Lambton, "The Regulation of the Waters of the Zayanda Rud," *BSOAS* 9 (1937–39): 665.

happens, Iskandar Beg was the author who dealt most directly with the problems of chronology. As a secretary (*munshi*) his duties included drafting letters, documents, diplomatic messages, and keeping financial accounts.[12] He wrote: "I ... conceived a desire to learn bookkeeping. ... I lost no opportunity of studying biographical and historical works, because I wanted to achieve success in that noble brand of learning."[13] Careful with his evidence, he only wrote about things which he had himself witnessed or about which he had consulted reliable sources.[14]

His *Tarikh-i Alam Ara-i Abbasi* (*The World Adorning History of Abbas)* was divided into two parts: the first, completed in 1616 and organized topically, gave abbreviated accounts of Muhammad and the Twelve Imams, the origins of the Safavid Sufi order, and the reigns of the first four Safavid rulers. The second, finished in 1629, was a detailed history of Abbas I. In his preface, Iskander Beg stated that he had decided to organize his material annalistically (i.e., year by year [*sal bi sal*]). There was, however, a problem.

> If, in the manner of historians [*ahl-i tarikh*], the Hijra year, whose beginning according to Arab usage is the first of Muharram, should be adopted, then most of the people of Iran would not understand. For among the Turks and Iranians, the beginning of the year is Nau Ruz-i Sultani, which is the first day of the world-adorning spring. When four seasons pass and another Nau Ruz comes, that is one year. The month of Muharram may fall anywhere in the Turki year [*sal-i Turki*]. Dates of great events [of the Hijra year] for which talented people provide a literary arrangement will vary with the Turki year.... Some will correspond to the previous year and some to the subsequent, i.e., one fewer or one more. Since this ignoble atom is under the obligation to eliminate and expunge ambiguities and problems from this work ... [and make] it acceptable to the comprehension of both the ordinary person and specialist alike, he has closed his eyes to that discrepancy ... and has settled on the Turki year which the general public better understands so that knowledge may increase among those who inquire after events.[15]

Despite his awareness of the difficulties, Iskandar Munshi was not able to completely escape chronological confusion. In fact, a brief look at several passages shows how difficult clarity could be, even for the most experienced and thoughtful historian.[16] Abbas I came to the throne on

[12] Sholeh Quinn, *Historical Writing During the Reign of Shah Abbas: Ideology and Legitimatcy in the Safavid Chronicles* (Salt Lake City: University of Utah Press, 2000), 22.
[13] Ibid., 58.
[14] Ibid.
[15] Iskandar Munshi, *Tarikh-e Alamara*, 2: 547–8.
[16] For a discussion see R. D. McChesney, "A Note on Iskandar Beg's Chronology," *Journal of Near Eastern Studies* 39 (1980): 53–63.

16 October 1587, and the heading of Iskandar Beg's first chapter read: "The Remaining Events of the Year of the Pig, the Year of Shah Abbas's Accession . . ."[17] Because the Year of the Pig ran from 21 March 1587 to 20 March 1588 and because the official accession date, in Safavid Iran as in Mughal India, was the first New Year's Day after the emperor's assumption of rule, this heading was accurate. Abbas's first year lasted only about five months – 16 October 1587–20 March 1588. Although the Hijra Era year of Abbas's accession went unremarked, Iskandar Beg's heading for the emperor's second year was "The Year of the Rat, the Second Year of Shah Abbas's Reign." The first line of this chapter began "In the Spring of 997 . . .," 997, however, was not the accurate Hijra equivalent and this error highlights the difficulties Safavid historians had in integrating the calendars of the solar and lunar eras. The Year of the Rat, the second year of Abbas's reign, and the first to encompass a full 365 days, ran from 21 March 1588 to 20 March 1589. Because Hijra 997 ran from 20 November 1588 to 10 November 1589, its spring would have begun 21 March 1589 not 1588. The Hijra spring at the beginning of Abbas's second year fell in 996 (2 December 1587–19 November 1588). Iskandar Beg's confusion probably arose from the Hijra date of the emperor's accession. When Abbas came to the throne (16 October 1587), Hijra 995 had only about six weeks to run – it ended on 1 December 1587. Thus, the bulk of Abbas's first year fell in Hijra 996 (2 December 1587–19 November 1588). As a result, it was only natural for the historian to have chosen 997 as the Hijra equivalent for beginning of Abbas's second year. And, in fact, this confusion has crept into the modern accounts: many authors giving 996/1588 instead of 995/1587 as the beginning of Abbas's reign.[18]

For the Safavid historian the problem of chronology arose not only in choosing the proper Turkish and Hijra Era equivalents but also in calculating the Hijra date of Nau Ruz. In Safavid Iran the first day of spring was a major occasion: It was the first day of the civil year, the anniversary of the emperor's accession, and the due date for agrarian taxes. Nevertheless, arriving at the exact Hijra date (day, month, year) for 1 Farvardin (21 March) was not easy. Although a skilled astrologer, working from the conversion tables of an up-to-date almanac, could in theory solve the

[17] Iskandar Munshi, *Tarikh-e Alamara*, 2: 551.

[18] While H. R. Roemer in the authoritative account in the *Cambridge History of Iran* got it right (6: 261, 350), several other authors have been confused. I. M. Lapidus, *History of Islamic Societies* (Cambridge: Cambridge University Press, 2002), 280 has 1588–1629 and Sholeh Quinn, *Historical Writing*, p. 3 has 996/1588–1039/1629.

problem, the chronicler might easily fall into error. An inexperienced astrologer, a flawed almanac, an inattentive copier – any of these could throw off the calculation. Given his awareness of the problem, it is ironic that Iskandar Munshi seems to have committed so many errors.[19]

The second historian was Mahmud b. Hidayat Allah, Afushtah-i Natanzi (b. 1531). The oldest one of the four, Natanzi was nearly sixty-eight years old when he wrote *Nuqawat al-Athar fi Dhikr al-Akhyar* (*The Choicest of Works in Remembrance of the Righteous*) in 1598.[20] Like Iskandar Munshi, Natanzi was not a full-time historian but, since he had access to imperial documents and correspondence, he must have had some position at court.[21] Natanzi's account, however, unlike Iskandar Munshi's, was limited in scope – not only to the Safavid dynasty but to the reigns of four rulers: Book One covered Tahmasp, Ismail II, and Muhammad Khudabanda while Book Two dealt with the initial eleven years of Abbas I.

Because of the leisurely pace of his narrative – he devoted more than 350 pages to eleven years – Natanzi seems to have decided to organize his material thematically rather than chronologically – a single section sometimes encompassing events of several years. His table of contents reveals the following chapter titles: Victory of Mushid Quli Khan over the Uzbeks, Reconquest of Mashhad, Abbas's Entry into Qazvin, and Abbas's Entry into Isfahan. Because in Book Two he was writing at roughly the same time as the events he was describing (i.e., he drafted his account of Abbas's first eleven years (1587–1598) during the period 1590–1598), he perhaps felt less need to include dates. Nevertheless, when he did employ the annal format for chapter headings, he had a different way of integrating the solar and lunar eras. For example, an early chapter was entitled: "Describing the Events of the Fourth Year After the Accession of Abbas; i.e., 14 Jumada I 998–24 Jumada I 999."[22] Although at first glance it appears as if Natanzi had chosen the liturgical era as his default format, a closer look suggests that the underlying historical chronology was in fact solar. First, although the months and years were lunar, the twelve months so delimited encompassed not the 355 days of the Hijra year (14 Jumada I 998–13 Jumada I 999) but rather the 365 days of the Turkish and Jalali era years (the extra

[19] McChesney, "A Note on Iskandar Beg's Chronology."

[20] Mahmud b. Hidayat Allah, Afushtah-i Natanzi, *Nuqavat al-Asrar*, ed. Ihsan Ishraqi (Tehran: BYNK Press, 1971); Quinn, *Historical Writing*, 20.

[21] R.D. McChesney, "Four Sources on Shah Abbas's Building of Isfahan," *Muqaranas* 5 (1988): 104–05; Quinn, *Historical Writing*, 20, 54–7.

[22] McChesney, "Four Sources," 107.

ten days of 14 Jumada I – 24 Jumada I). Second, the year began not on the first day of the liturgical year (1 Muharram) but rather on 1 Farvardin (21 March). Finally, Natanzi also adopted the Iranian custom of moving the official accession date to the nearest Nau Ruz. Thus, like Iskandar Munshi, he had Abbas's first year end on 20 March 1588, and, as a result, his fourth year ran from 21 March 1590 to 20 March 1591. Although Natanzi was sparing of dates, he must have had access to a skilled astrologer. Not only were his Hijra equivalents for Nau Ruz accurate, but he occasionally provided dates for important events, giving the solar and lunar versions. Thus, Abbas's father, Muhammad Khudabanda, ascended the throne on 12 Shawwal 985 in the Hijra Era and 1 Urdubihisht 498 in the Jalali.[23]

The third historian was Jalal al-Din Munajjim Yazdi (d. 1619). Occupying the office of chief imperial astrologer (*munajjim bashi*), he was a great favorite of the young emperor. In 1609 Abbas dispatched him to Hulaghu's observatory in Maragha for study and observation. After his death the emperor appointed his son, Mulla Kamal (author of *Tarikh-i Mulla Kamal or History of Mulla Kamal*) and *Zubdat al-Tawarikh (The Best of the Histories*), to his father's post. Mulla Kamal remained chief astrologer throughout the reign of Shah Safi, Abbas's successor.[24]

Because of his training and experience, Jalal al-Din had the surest command of chronology. According to the historian Abd al-Husayn Khatunabadi among the most important sources were the "... orders of the astrologers [*dastur-i munajjiman*]."[25] Jalal al-Din's history – *Tarikh-i Abbasi* (*History of Abbas*) *or Ruznama-i Mulla Jalal* (*Mulla Jalal's Almanac*) – began, like the *Akbar Nama*, with an astrological prologue: the interpretation of Abbas's horoscope at birth (nativity). Like Natanzi, his chronicle covered a relatively short span of Safavid history. It began with the death of Shah Tahmasp (1576) and continued until 1611–1612, the twenty-fifth year of Abbas's reign. Because of his profession Jalal al-Din included more dates – with greater detail and more equivalents – than did the other three chroniclers. Although he divided his annal according to the Hijra Era, the first line of each chapter included a solar equivalent. Thus, "A Description of the Events of 1000" began "... according to the Tushqan [Rabbit] Year he turned on Sih Shambah [Tuesday] or 4 Muharram toward

[23] Natanzi, *Nuqavat*, 64.
[24] Ali Asghar Mossadegh, "La Famille Monajjem Yazdi," *Studia Iranica* 16 (1987): 125–9.
[25] Quinn, *Historical Writing*, 110.

Isfahan."[26] And the first line of his Hijra 1009 chapter was "The first of Muharram of this year was equivalent to the 24th Tir [Fourth Month] of Sichual [Mouse] Year."[27] In contrast to Iskandar Munshi and Natanzi, however, the significant New Year was not the Zoroastrian 1 Farvardin (21 March) but the liturgical 1 Muharram.

The final historian was Mirza Beg b. Hasan Husain Junabadi. His *Rauzat al-Safawiya* (*The Garden of the Safavids*) was devoted entirely to the Safavid dynasty – from its founding in 1501 by Ismail to the death in 1629 of Abbas I. Like the other three men Junabadi was a contemporary of Abbas, beginning his work in 1617 and finishing sometime in the 1630s. Nothing is known of his occupation but he was probably a minor court functionary as he seems to have witnessed many of the events he described. Although Junabadi included a Hijra date in the headings of a few early chapters,[28] a typical chapter title was prolix and awkward, without any chronological marker whatsoever. For example, the chapter on Abbas's founding of his new capital, one of the most important passages in the entire work, began: "A Description of the Creation of the Buildings and Structures of Dar al-Saltanat-i Isfahan and the Rejuvenation of the Suqs [Covered Bazars] and the Creation of the Naqsh-i Jahan Maidan and the Great Buildings and Gates and the Laying out of the Avenue and the Chahar Baghs [Garden Palaces] of the City of Isfahan."[29] About halfway through the chapter, however, a date appeared: "In 1012 at a well-chosen hour a new maidan [piazza] was laid out in the Naqsh-i Jahan area."[30]

MUGHAL EMPIRE

Fiscal

India, like Iran, had a long history of solar and lunisolar calendars and eras. Of the three subcontinental systems of historical accounting, the most ancient was the Mahayuga or Great Era. Encompassing a total of 4,320,000 years, this era was further subdivided into four shorter *yugas*. The first was the Krita or Golden Yuga (1,728,000 years), the second was the Treta or Silver Yuga (1,296,000 years), and the third was the Dvapara

[26] Ibid., 113.
[27] Ibid., 200.
[28] Mira Beg. B. Hasan Husaini Junabadi, "Rauzat al-Safaviyah," London, British Library Persian Manuscript Collection, Or. 3388, fol. 268a, 290a.
[29] Ibid., fol. 313b.
[30] Ibid., fol. 314b.

or Bronze Yuga (864,000 years). The last and current era, the Kali Yuga or Black Era, had begun at midnight 17/18 February 3102 BCE and would continue for 432,000 years. According to Hindu mythology, the inhabitants of the subcontinent had deteriorated over the four eras – declining in physical strength, life expectancy, and moral and intellectual stature. Although the Kali Yuga was described in most astronomical treatises and almanacs, it was not generally found in local histories or records. The Vikramaditya Samvat (Vikramditya Era), by contrast, was in everyday use. Said to have been established by King Vikramaditya, the ruler of Ujjain, to commemorate his victory over the Sakas, its epoch was 14 October 58 BCE. The Saka Samvat (Saka Era), the third era, had been established by the Indo-Scythians or Sakas, a branch of the larger Indo-European Saka or Scythian group of tribes who entered Afghanistan and northern India during the period 175–150 BCE. Its epoch was 3 March 78 CE, and it was employed in many Indian states.[31]

Well before the Mughals, Muslim revenue collectors and administrators had been aware of the inadequacies of a lunar era in an agrarian state. In 1342 the Indian mathematician and astronomer/astrologer Hajji Abdul Hamid Ghaznavi advised:

> ... the difference between the solar and lunar year is 10 13/15 days. Accordingly, 33 solar years are equal to 34 lunar years. If the [tax collector] does not know this, he will collect revenue from peasants for 68 half years while the actual number of years will be 66. This amounts to extortion and exploitation of the poor ...[32]

While Babur was aware of the Vikramaditya Samvat and the Iranian eras,[33] neither he nor his son was able to achieve the military and institutional stability that would have allowed the introduction of a new system of revenue administration and a new era. It was Babur's grandson Akbar who introduced the first solar era – a taxation system called the Fasli or Harvest Era. Although we do not know exactly when this was inaugurated, it probably coincided with the ruler's efforts in the 1570s to reform the Mughal system of land revenue administration. Like the Kharaji, the Fasli Era was tied to the Hijra. Its first year (963 AF) matched the first year of Akbar's reign (963 AH/1556 CE). In South India the Bahmani Sultanate

[31] Sukhvir Singh Gahlot, *Indian Calendars* (Jodhpur, Rajasthan: Rajasthan Sahitya Mandir, 1980); Alexander Cunningham, *Book of Indian Eras*, reprint ed. (Varanasi: Indological Book House, 1970); Pillar, *Indian Chronology*.

[32] A. Rahman, ed., *History of Indian Science, Technology, and Culture, ad. 1000–1800* (New Delhi: Oxford University Press, 1999), 56–7.

[33] Sayili, *Observatories*, 264; Abu al-Fazl, *Akbar Nama*, 1: 121, fn. 2.

had established a solar taxation era some two hundred years earlier. The initial year of the Deccani Shuhur (Solar) Era (744) also matched the Hijra Era year (744 AH/1344 CE).[34] The Fasli Era was used almost exclusively in the revenue and accounting departments – for land grants, gifts, revenue assignments, payments of wages and pensions, and remissions or exemptions from taxes. It was almost never seen in court documents, inscriptions, or histories.[35] Because most of the cultivators in Mughal India were Hindus or converted Muslims, the Mughals collected the agrarian taxes according to the indigenous calendar: the spring taxes were due after Holi and the fall taxes after Dussehra.[36]

Since the Fasli Era suffered from the same deficiency as the Kharaji – that is, it had to be advanced every thirty-two years to keep it in harmony with the Hijra Era – Akbar's men employed a second solar era. The Turkish Twelve-Year Animal Era – Sanawat-i Turki (Turkish Era) or Duwazdah Sal-i Turki (Twelve-Year Turkish Era) – was an import from Safavid Iran and was used in financial and administrative documents. The animal year was typically preceded by a mention of the harvest – either fall (*kharif*) or spring (*rabi*). A revenue grant, for example, would begin "*az ibtida-i kharif-i yunt yil* (from the beginning of the fall harvest of the horse year)." It was followed by the Hijra year, the regnal year, and after 1584 the Ilahi year month and day. Documents containing the animal era and dated as early as 1558, just two years after Akbar's accession, have been discovered.[37] In keeping with his "lasting reconciliation" ethic of religious and cultural tolerance Akbar prohibited (at least for the members of his Tauhid-i Ilahi) the injury or killing of the eponymous animal during the twelve years of the Turkish calendar. In the mouse year no mice should be injured, in the leopard year leopards should not be captured, in the snake year serpents should not be killed, in the cock year birds should not be killed or used to fight, and so on.[38]

Despite all of the chronological systems available – the Islamic, the Indic, and the Iranian – Akbar decided midway through his reign to introduce a new era. Because of the sanctity of the Hijra, such an innovation was extremely rare. In fact, it had been more than five hundred years

[34] G. H. Kare, "The Turkish Duodenary Cycle and Its Use by the Mughals in India," *Islamic Culture* 26 (1952), 64–74; *Encyclopedia Britannica*, s.v. "Fasli Era."

[35] Gahlot, *Indian Calendars*, xv–xvi; Cunningham, *Book of Indian Eras*, 82; Pillar, *Indian Chronology*, 44–5.

[36] Abu al-Fazl, *Ain* 2: 48.

[37] Kare, "The Turkish Duodenary Cycle;" Abu al-Fazl, *Ain* 2: 21–2.

[38] Nizami, *Akbar and Religion*, 140–1.

since the creation of a new public chronology, one widely employed in chronicles, proclamations, orders, and ceremonies.[39] Thus, Akbar's introduction of the *Tarikh-i Ilahi* (Divine Era) in 1584 CE, with an epoch dated to the Vernal Equinox nearest his accession, 21 March 1556 CE, was both revolutionary and controversial.

To gain a better understanding of the era itself and of Akbar's reasons for introducing it, we must take a brief look at the man in charge of its design, Mir Fathullah Shirazi (d. 1589).[40] A polymath and, as his name indicates, an emigrant from the famous city of poets, Shirazi was charming, witty, and immensely learned. He completely captivated the emperor at a critical point in his remaking of the Mughal state.

Shirazi was part of a larger stream of talented and ambitious men who had migrated to the Muslim kingdoms of the Deccan over the two hundred years from about 1350 to 1550. Merchants, scholars, physicians, poets, painters, soldiers, and administrators – these men traveled through the Persian Gulf ports of Hormuz and Gombroon (later Bandar Abbas), drawn by the lucrative career opportunities in the rich, populous states of South India.[41] These Iranis, however, must be distinguished from a later group of immigrants who arrived overland via the northern cities of Qandahar and Kabul. Leaving the Shiite empire of the Safavids, these men – representing the same range of occupations – were attracted by the wealth and scale of the north Indian empire of the Mughals. Shirazi, in addition to his other accomplishments, was one of the first to bridge these two Persianate Indian worlds – the South and the North. He began in the Deccan, at the tail end of the South Indian surge of immigration, and ended his career at Akbar's court.

Although very little is known of Mir Fathullah's early life in Shiraz, there is no doubt that he possessed a powerful intellect and had acquired an impressive store of knowledge. Although he was skilled in the transmitted or traditional sciences (Arabic grammar, Quranic interpretation, traditions of the prophet, rhetoric, and jurisprudence), it was in the rational sciences (logic, philosophy, mathematics, astronomy, astrology, and medicine) that he made his reputation – both in Iran and in India. He attended the Mansuriyyah Madrasa (birthplace of the famous philosophical School of Shiraz) and studied under Mir Ghiyath al-Din Mansur

[39] The Jalali Era, epoch 21 March 1079 CE.

[40] Abu al-Fazl, *Akbar Nama*, 2: 15–18.

[41] Richard Eaton, *A Social History of the Deccan, 1300–1761: Eight Indian Lives* (Cambridge: Cambridge University Press, 2005), 59–80.

Dashtaki (1461–1542), the foremost philosopher of sixteenth-century Islam.[42] The Madrasa had been founded in 1478 by Mir Ghiyath al-Din's father, Sadr al-Din Dashtaki (1424–1497), and had quickly acquired a reputation for excellence in the rational sciences.[43] Sadr al-Din named the madrasa after his son, and although he himself wrote more than twenty works of philosophy and theology, he was overshadowed by the two principal figures of the school – his son and Jalal al-Din Dawani (1427–1502). Dawani wrote a good deal on a variety of topics but his fame, both in his day and ours, rested on his ethical treatise – *Akhlaq-i Jalali*. Based on the *Akhlaq-i Nasiri* of Nasir al-Din Tusi, Dawani's work commanded a wide readership in the Mughal and Ottoman empires as well as in the Safavid.

For Shirazi, however, the principal figure in the School of Shiraz was Mir Ghiyath al-Din. Unlike his father and Dawani, Mir Ghiyath al-Din made his mark in the rational rather than in the traditional sciences. His works spanned an impressive range, from theological, mystical, and Quranic studies to treatises on medicine, mathematics, astronomy, and astrology. He also composed an essay on ethics (*Akhlaq-i Mansuri*), modeled after the treatises of Tusi and Dawani. After Ismail's conquest of Shiraz in 1508 he confirmed Mir Ghiyath as head of the Mansurriyah (probably assigning him an additional stipend) and appointed him to the high office of Sadr (Chief Religious Officer). Over the next several years the shah consulted him on a variety of topics – the proper time and place for initiating important activities and the treatment of fevers, for example. Mir Ghiyath al-Din followed Ismail to Tabriz and in 1521 was dispatched to Maragha to oversee the repair of Tusi's observatiory and to update the *Zij-i Ilkhani*. After Ismail's death in 1524 Dashtaki remained an important councilor for the young Tahmasp – constructing, for example, a talisman to ensure victory in an early battle. With time, however, Tahmasp came to depend more and more on the Shiite ulama. A conflict erupted in 1530 when Tahmasp ordered the leaders of the two camps – Mir Ghiyath al-Din of the rational sciences and Sheikh Ali Karaki of the religious sciences – to reconcile their differences. When Dashtaki refused to meet with Karaki (because, he said, the astrological signs were inauspicious), Tahmasp

42 Sajjad H. Rizvi, "Mir Damad and the Debate on Hudut-i Dahri in India," in Denis Herman Denis and Fabrizio Speziale, eds., *Muslim Cultures in the Indo-Iranian World During the Early-Modern and Modern Periods* (Berlin: Klaus Schwarz Verlag, 2010), 449.

43 Seyyed Hossein Nasr, *Islamic Philosophy from Its Origin to the Present* (Albany: State University of New York Press, 2006), 193–208.

became uneasy. And when the Mir maintained that astronomical calculation, not Shiite tradition, was the proper technique for determining the qibla for new mosques, Tahmasp dismissed him (in 1531) from his post as Sadr.[44] He returned to Shiraz and resumed his headship of the Madrasa, teaching until his death in 1542.[45]

It was during these last ten years that the young Mir Fathullah began his studies at the Mansurriyah, mastering Dawani's philosophical system and learning medicine, mathematics, astronomy, and astrology at the feet of Mir Ghiyas al-Din and others.[46] A precocious and hard working student, he soon became the leading intellectual light of the School of Shiraz, probably succeeding Mir Ghiyath al-Din as head of the Madrasa. His reputation was based on his scholarly output: a five-volume commentary on the Quran, commentaries on the philosophical works of Tusi and Dawani, a Persian translation of Ibn Sina's *Canon of Medicine*, and many other compositions in astronomy, astrology, and mathematics.[47] Given his controversial career in India, it is interesting to note that he was something of an intellectual rebel in Iran as well. He was a follower of Azar Kayvan (1533–1618), a heterodox visionary[48] who looked back to the pre-Islamic Iranian past and preached a belief system that combined Islamic and Zoroastrian (Mazdean) elements. Kayvan attracted a large group of disciples (called Azaris or Kayvanis), but the opposition of the Safavid ulama caused him to flee Shiraz for India in the 1570s.[49] Because of his fame as a scholar and teacher, Shirazi attracted a great number of talented students during his more than thirty years at the Mansurriyah.[50] Among the most successful were Afzal Khan, Wazir of Ali Adil Shah I of Bijapur (1558–1580); Farid al-Din Ibrahim Dihlavi, compiler of the *Zij-i Shahjahani*; and Abd al-Rahim Khan-i Khanan, the cultured great amir of Akbar's court.

44 Kathryn Babayan, "The Cosmological Order of things in Early Modern Safavid Iran," in Bagci and Farhad, eds., *Falnama*, 252–53. For a further discussion of Karaki and the qibla controversy see Andrew Newman, "Towards a Reconsideration of the 'Isfahan School of Philosophy': Shaykh Baha'i and the Role of the Safawid Ulama," *Studia Iranica* 15 (1986): 181–5.

45 Nasr, *Islamic Philosophy*, 199–202.

46 Francis Robinson, "Ottomans-Safavids-Mughals: Shared Knowledge and Connective Systems," *Journal of Islamic Studies* 8 (1997): 158–9.

47 M. A. Alvi and A. Rahman, *Fathullah Shirazi: A Sixteenth Century Indian Scientist* (New Delhi: National Institute of Sciences of India, 1968), 23–4.

48 Badaoni, *Muntakhab*, 3: 216.

49 Mohamad Tavakoli-Targhi, "Contested Memories of Pre-Islamic Iran," *Medieval History Journal* 2 (1999): 245–76.

50 Badaoni, *Muntakhab*, 3: 216.

Shirazi first came to India at the invitation of his student, the ambitious Afzal Khan. Immigrating to Bijapur in the early part of Adil Shah's reign, Afzal Khan founded a school. His fame as a teacher soon brought him to the ruler's attention, and he entered the Bijapur administration, rising quickly to become prime minister in 1564.[51] Afzal Khan had not, however, forgotten his illustrious mentor, and in the mid-1570s he convinced Ali Adil Shah to invite Shirazi to court, sending him a large sum of money to smooth the transition. The news of Shirazi's arrival reached Akbar in Fathpur Sikri, and, while the Mughal could not convince the Bijapur sultan to share his treasure, Akbar's luck changed in 1580 when Adil Shah died. Under his successor, Ibrahim Adil Shah II (1580–1627), the high-ranking Shiite immigrants became suddenly unwelcome. Shirazi was imprisoned but, at Akbar's order (8 July 1582), he was released and escorted to the Mughal court,[52] arriving in Fathpur Sikri in early 1583.[53] Received with great honor, he was given the daughter of an important amir in marriage.[54]

Although Shirazi's career at the Mughal court was short (about six years), he had a profound impact on Akbar and the fledgling Mughal state. His work in mathematics, astronomy, astrology, engineering, and administration was undoubtedly important but his most valuable contribution was his introduction of the School of Shiraz. On the Mir's advice Akbar made the rational sciences (especially the works of Dawani and Mir Ghiyath al-Din) obligatory in the curricula of Mughal madrasas, both Sunni and Shiite.[55] And it was Mir Fathullah's intellectual descendants who founded the famous Farangi Mahall in Lucknow, which quickly became the most important center of Islamic philosophy in India. Its curriculum (the Dars-i Nizami) included a heavy dose of the rational sciences.[56]

Shirazi's training in Iran as a *munajjim* and his experience – preparing almanacs, casting horoscopes, and producing talismans and incantations – made him especially qualified to craft and implement Akbar's new solar

[51] S. A. A. Rizvi, *The Socio-Intellectual History of Isna' Ahsari Shi'is in India*, 2 vols. (Canberra: Ma'rifat Publishing, 1986), 1: 271–2.

[52] Abu al-Fazl, *Akbar Nama*, 3: 578.

[53] Ibid., 3: 593.

[54] Rizvi, *Socio-Intellectual*, 222–3.

[55] For a discussion see Ibid., 346; Nasr, *Islamic Philosophy*, 204–7; Sajjad Rizvi, "Mir Damad," 449; and G. A. Lipton, "Muhibb Allah Ilahabadi's Taswiya Contextualized," in Dennis, *Muslim Cultures*, 486.

[56] Nasir, *Islamic Philosophy*, 204–07; Robinson, "Shared Knowledge."

era – the *Tarikh-i Ilahi* or Divine Era.[57] He supplied Abu al-Fazl with the correct Ilahi years and Zoroastrian days and months for the pre-1584 chapters of the *Akbar Nama* and prepared one of the nativities (birth horoscopes) reproduced by Abu al-Fazl in the first chapters of his history.[58] He also translated Ulugh Beg's *Zij-i Sultani* into Sanskrit.[59]

During his first months in Fathpur Sikri Mir Fathullah designed and constructed several mechanical devices. In Safavid Iran, in contrast to Mughal India, it had long been the tradition for a cultured man to master a craft, displaying his skill as an artisan. Shah Abbas, for example, had boasted of his handiwork.[60] Shirazi composed a treatise on mechanics and hydrology entitled *Kalid-i Danish* (*Key to Knowledge*),[61] and at the Nau Ruz festivities of 1584 he exhibited several of his creations: a self-grinding millstone, a distorting mirror, and a twelve-barrel cannon.[62]

The Mir also knew a great deal about land revenue administration. In 1580 Akbar's men published the Ain-i Dah Salah (Ten Year Regulations), and in 1582 Raja Todar Mal was appointed Wazir (Head of the Revenue Office). Because Todar Mal had adopted Iranian administrative procedures and because Persian was the language of administration, Akbar in 1584 appointed Shirazi Amin al-Mulk (Head Administrator) and asked him to look over the new regulations.[63] At the emperor's order his list of twenty improvements was implemented.[64] To facilitate the process of monetization the Iranian also drafted a detailed guide for the valuation and exchange of worn coins.[65] In 1585 Shirazi was appointed Sadr al-Sudur (Head of the Charitable Grants Office) and undertook a rationalization of the entire system – reexamining the documents of current holders and issuing new rules.[66] In 1586 he was given a higher rank and

57 Khwajah Nizamuddin Ahmad, *Tabaqat-I-Akbari*, trans. Brajendranath De, revised and ed. Baini Prasad, 3 vols. (Calcutta: Asiatic Society of Bengal), 2: 684–5; Badoni, *Muntakhab*, 3: 216.

58 Abu al-Fazl, *Akbar Nama*, 1: 125–7.

59 Ghori, "Development of Zij Literature," 392.

60 Mehdi Keyvani, *Artisans and Guild Life in the Later Safavid Period* (Berlin: Klaus Schwarz Verlag, 1982), 41–2.

61 S. M. Razallah Ansar, "On the Transmission of Islamic Astronomy to Medieval India," in *The Tradition of Astronomy in India: Jyotihsastra*, 365, fn. 7.

62 Khan, *Maathir*, 1: 544–5; Ahmad, *Tabaqat*, 2; 684–5; Rizvi, *Socio-Intellectual*, 2: 198.

63 Badaoni, *Muntakhab*, 2: 325; Khan, *Maathir*, 1: 543; Ahmad, *Tabaqat*, 2: 684–5.

64 Abu al-Fazl, *Ain* 3: 457–9.

65 Ibid., 1: 34–6.

66 Rizvi, *Socio-Intellectual*, 1: 225.

the title of Azad al-Daulah and dispatched on diplomatic missions to the Rajah of Khandesh and the governor of Bengal (Shahbaz Khan).[67]

In addition to his practical skills, he was a powerful intellectual presence – a member of Akbar's Sufi-like imperial order (Tauhid-i Ilahi) and a major contributor to the philosophical and theological discussions in the Ibadat Khana. To the horror of Abd al-Qadir Badauni and the other conservative ulama, he was an unapologetic Shiite. As a member of the Irani faction at court, he expounded the Safavid doctrines concerning the Mahdi and openly practiced Imami rituals. His heterodox beliefs and practices (acquired from Azar Kayvan) also contributed to Akbar's interest in Zoroastrianism, manifested in his reverence for the sun and his reported worship of fire.[68]

In May of 1589 Shirazi accompanied Akbar on a visit to Kashmir. He fell ill but, being a doctor himself, refused the help of the chief imperial physician. He died and was buried in Srinagar. Akbar was devastated.

> The Mir was at once our *vakil* (advisor), philosopher, physician, and astrologer. Who can comprehend the extent of our sorrow? If he had fallen into the hands of the Franks [Europeans] and they had wanted the whole of my treasures in exchange for him, ... I should have thought that splendid jewel to be very cheaply purchased.[69]

In the *Akbar Nama* the primary reason given for the introduction of the new solar era (the *Tarikh-i Ilahi*) was practicality. Dates are important, the historian wrote, to fix events, and thus various eras have been established, each beginning with a great event. Although Akbar's accession (14 February 1556) was a world-shattering event, worthy of inaugurating a new era, the emperor decided to shift its epoch five weeks forward, to the next Vernal Equinox (21 March 1556). In addition to following Iranian practice in calculating the beginning of each reign, the new era also employed the solar Zoroastrian calendar of the Jalali Era, with the Iranian day and month names. Akbar also established a twelve-year cycle, undoubtedly inspired by the Turkish Twelve-Year Animal Era. Thus, each Divine Era year was named after one of the twelve Zoroastrian months – Year One was Farvardin (Tarikh-i) Ilahi, Year Two Urdubihisht (Tarikh-i) Ilahi, and so on.

In chapter four of the *Akbar Nama* Abu al-Fazl reproduced the text of the *firman* (order) Akbar sent to his officeholders and officials on the

[67] Abu al-Fazl, *Akbar Nama*, 3: 593; Ahmad, *Tabaqat*, 2: 684–5.
[68] Robinson, "Ottomans-Safavids-Mughals," 158–9.
[69] Khan, *Maathir-ul-Umara*, 1: 544.

implementation of the new era. In the *firman*, however, the new dating system was not the emperor's idea. Rather, in a brilliant piece of rhetorical strategy, Abu al-Fazl gave the argument for practicality to several high-ranking officeholders, placing their "fictitious" petition within the body of the order itself. The various eras, the petition argued, had become cumbersome and awkward: The Indian Era (Vikramaditya) now exceeded 1500, the Era of Alexander was more than 1000, and the Hijra and Yazdegird Eras were nearly 1000. Thus, "The writing and speaking of such eras in conversation and in business [had become] very difficult for men of the world, and especially so for the commonalty who are the centre of business."[70] If Akbar were to establish a new era, it would be "a mark of thanksgiving and a fulfillment of the wishes of mankind."[71]

The purpose, it would appear, of this invented petition was to defuse the opposition of pious Muslims. The change was a necessary revision of administrative and commercial practice (at the request of overworked merchants and officials), not an attack on Islamic tradition. In fact, the petition continued, there was a precedent: the Seljuq Malik Shah had introduced the Jalali Era in 1079 CE (for reasons of convenience and accuracy) and it had become accepted throughout the Islamic world – in the almanacs (*taqvims*) of Arabia, Turkey, Transoxiana, Khurasan, and Iraq. Akbar intended no disrespect to Islam.

Akbar's *firman* concluded by ordering imperial officials to begin using the new era as of 21 March 1584. Islamic astronomers and astrologers were instructed to enter the Divine Era into their almanacs alongside the Arabic (Hijra), Rumi (Christian), Persian (Yazdegird), and Jalali eras, and Indic astrologers were urged to substitute the new era for the Vikramaditya. Finally, a new series of solar festivals, based on the Zoroastrian calendar, was introduced. The only ones which took root, however, were the first two – which celebrated the beginning and end of Nau Ruz.

Abu al-Fazl's second discussion of the new era was found in the Introduction to Book Three of the *Ain-i Akbari*. In his translation H. Blochmann added the title "Divine Era." This was doubly erroneous: Not only had Abu al-Fazl not included a title in the original Persian but his subject matter was much more comprehensive than the invented title would suggest. The introduction began with a list of eighty-six astronomical treatises – including those of Ptolemy, Zoroaster, al-Biruni,

[70] Abu al-Fazl, *Akbar Nama*, 2: 21.
[71] Ibid., 2: 22.

Mashaallah, Abu Mashar, Tusi, and Ulugh Beg. It continued with an account of the Indic eras and the current Islamic eras before finally turning to the Divine Era.[72] "His Majesty," Abu al-Fazl wrote, "had long desired to introduce a new computation of years and months throughout the fair regions of Hindustan in order that perplexity might give place to easiness ..."[73]

Here Abu al-Fazl referred to the familiar problems raised by a lunar era in an agrarian state. Of the Hijra Era in the Ilkhanid state he wrote:

> ... the road was opened to grievous oppression, because 31 lunar years are equal to only 30 solar years and great loss occurred to the agriculturists, as the revenue was taken on the lunar years and the harvest depended on the solar ...[74]

Dating

In India there was a widespread interest in chronology, and Mughal historians, like their Safavid counterparts, had a variety of eras to choose from. In one of the early Sanskrit histories (Puranas, ca. 500) the ruler is instructed to appoint a chronologer (*samvatsara*): "He who knows the entire year."[75] And Naqib Khan (d. 1573), the historian who tutored the young Akbar, "... could give offhand the date of any significant event in Islamic history (battles, births, dates, reigns). ... 'No one has equaled him in the knowledge of chronology.'"[76]

The first history written under the Mughals reflected both an interest in chronology and an awareness of the impending millennium. In 990/1582 – the year of the Grand Conjunction of Jupiter and Saturn – Akbar ordered the compilation of a universal history, the *Tarikh-i Alfi* (*Millennium History*). Not only was the millennium approaching (1000 AH/1591–92 CE), but, according to Akbar, it had been exactly one thousand (lunar) years since the prophet's first revelation (i.e., 612).[77] In the same year the emperor also issued a new set of coins stamped with the word *alf* (one thousand).

[72] Abu al-Fazl, *Ain* 2: 1–32.
[73] Ibid., 2: 29.
[74] Ibid., 2: 28.
[75] Ronald Inden, "Imperial Puranas," in Ronald Inden, Jonathan Waters, and Daud Ali, eds., *Querying the Medieval: Texts and the History of Practices in South Asia* (New York: Oxford University Press, 1999), 40.
[76] Khan, *Maathir-ul-Umara*, 2:1, 381–2.
[77] Badaoni, *Muntakhab*, 2: 310.

The Millennium History was an annal, to be written by a committee of seven scholars: each man was to take a year, and they were to alternate until the story of Islam was complete. The list of names is significant: Year One was Naqib Khan, son of Akbar's tutor, and an eminent chronologer himself; Year Two was Mir Fathullah Shirazi; Year Six was Nizam al-din Ahmad (author of *Tabaqat-i Akbari*); and Year Seven was Abd al-Qadir Badauni (author of *Muntakhab al-Tawarikh*). In a preview of his burgeoning interest in chronology (the *Tarikh-i Ilahi* would be implemented two years later), Akbar decreed that the new history should have a new era – its epoch would be 632 (the year of the prophet's death) rather than 622 (the year of his migration from Mecca to Medina). After the committee had covered the first thirty-five years, Akbar became frustrated by the slow pace and decided to adopt a more conventional method. He appointed two new historians: Mulla Ahmad brought the narrative down to the reign of Ghazan Khan (1271–1304) (vol. 2), and Jafar Beg Asaf Khan continued the story to the present – 997/1588–1589 (vol. 3). Although Abu al-Fazl is said to have contributed a preface, no trace of it has survived. While largely ignored today, the work was not without value: the last section of the third volume was the first official history of Akbar, and Nizam al-Din relied on it for his narrative of Humayun's reign.[78]

Nizam al-Din Ahmad's (1551–1621) *Tabaqat-i Akbari* was the first general history of Islamic India and the first serious account of Akbar's reign.[79] Begun in the late 1580s and finished in 1593–1594, it covered the history of Indian Muslims from 987/88 until the latter years of Akbar's reign. In addition to the Millennium History, Nizam al-Din also consulted drafts of the *Akbar Nama*. His history, like Abu al-Fazl's, was divided according to the recently adopted *Tarikh-i Ilahi*, but he often included Hijra Era equivalents. For example, his account of Akbar's first year began:

> Let it not remain concealed that the Ilahi year is the name of the solar and correct year and its beginning has been fixed from Nau Ruz. The beginning is Monday, 27 Rabi II 963 (10 March 1556).[80]

For most of his narrative, however, Nizam al-Din employed the Hijra rather than the Ilahi names for the days, months, and years. For example,

[78] Rizvi, *Religious and Intellectual*, 253–62.
[79] Ibid., 277–8; Khan, *Maathir-ul-Umara*, 2:1: 396.
[80] Ahmad, *Tabaqat*, 2: 209.

"... on the 17th Zu al-Hijjah of the sixth year of the Ilahi era corresponding to 969 AH."[81] By the time he reached the end of his account, revisiting the tumultuous events of the Fathpur Sikri years (1571–1585), he had begun to include more Ilahi days and months. In the penultimate year of his history, for example, he described an incident which occurred "... on 24 Amardad in the 38th year of the Ilahi Era corresponding to 14 Zu al-Qada 1001."[82]

Because Nizam al-Din was the first of Akbar's historians to employ the *Tarikh-i Ilahi* and because the Mughal astrologers had not had a great deal of experience with the new era, he made a number of mistakes in translating dates from one era to another. A common error was pointed out by his close friend, Abd al-Qadir Badauni. Nizam al-Din had forgotten, Badauni wrote, to take "... account of the intercalated days, which every three years makes a difference of one month, there is a difference in each cycle of a whole year, between the solar and lunar years."[83] The problem, for Nizam al-Din and later even for Badauni, was the same one that had bedeviled earlier historians and record keepers employing the Kharaji, Fasli, and Maali taxation eras. Although solar and lunar era years remained synchronized for thirty-two years, the eleven-day difference between the two meant that in the thirty-third year the numbering sequence would diverge. And, in fact, we find that Nizam al-Din's error occurred at about this point in his chronicle – after his account of Akbar's thirtieth year.[84]

Abd al-Qadir Badauni (1540–1615) was, as we have seen, much more conversant with issues of chronology. He had been a student of Abu al-Fazl's father, Sheikh Mubarak, and had been introduced to Akbar in 1574. A mature intellect and a persuasive speaker, he made a powerful first impression – especially in the debates in the Ibadat Khana. He was, however, temperamentally and intellectually unsuited to court life. After several unsuccessful efforts at promotion and a number of unauthorized absences, he returned to imperial service, remaining a minor functionary until the end of his life. He worked primarily in the translation bureau, helping to turn the Hindu religious classics (the Ramayana and Mahabharata) into Persian. He was also a historian. As one of the initial authors of the Millennium History, he was Akbar's choice in

81 Ibid., 2: 257.
82 Ibid., 2: 649.
83 Badaoni, *Muntakhab*, 2: 353.
84 Ibid.

1592 to revise the entire work, correcting the complicated and often erroneous chronology. About a year earlier, clearly underemployed, he began work on two other books: a philosophical/theological treatise entitled *Najat al-Rashid* and *Muntakhab al-Tawarikh*, the history by which he is best known.

Completed in 1595, *Muntakhab al-Tawarikh* was a history of Muslim rule in India. It comprised three volumes: vol. 1, the beginnings through the reign of Humayun; vol. 2, the reign of Akbar; and vol. 3, biographies of famous men. Although Badauni had Nizam al-Din's *Tabaqat* in front of him, his account had a completely different slant. Shocked by the religious and cultural diversity of Akbar's new order, he filled the pages of his history with disapproval and condemnation – much of it inaccurate or unfair. His history, like those of the other two, was an annal but he organized his chapters according to the Hijra rather than the Ilahi era – another example of his opposition to Akbar's new order. Nevertheless, he had mastered the intricacies of the new solar chronology and the techniques of converting dates from one era to the other. For example, in most of his chapters he included the Hijra day, month, and year of the first day of the Ilahi year (Nau Ruz), and he would occasionally supply the exact *Tarikh-i Ilahi* date (day, month, and year).[85] In addition to rectifying the chronological chaos of the *Tarikh-i Alfi* and correcting Nizam al-Din's errors, he also recorded the efforts of Akbar's men to fix the dates of Hindu festivals – according to the ancient eras rather than to the more modern Vikramaditya.[86]

The best-known and most famous history of Akbar's reign, however, was the *Akbar Nama*. A massive three-volume work, commissioned by the emperor and composed by Abu al-Fazl (1551–1602), it was begun in 1589–1590, after the completion of the *Tarikh-i Alfi*, and mostly finished by 1597–1598. The fact that Asaf Khan, the author of the Mughal section of the Millennium History, was not chosen as the author of the official history suggests that Akbar was not completely satisfied with that first account of his dynasty. Although the *Akbar Nama* was roughly contemporaneous with the works of both Nizam al-Din and Badauni, it was so huge that both men were able to examine early drafts while completing their own histories. Not himself a *munajjim* but very interested in chronology, Abu al-Fazl drew on the talents of the imperial experts – Mulla Chand and especially Mir Fathullah Shirazi. In fact, it appears that the

[85] See, for example, that for 990/1582. Ibid., 2: 310–12.

[86] Ibid., 2: 367–8.

historian composed his first two versions using the Hijra Era, and only later did Shirazi help him replace the lunar era dates with their solar era equivalents.[87] Even before he introduced the *Tarikh-i Ilahi* dates into his narrative, however, Abu al-Fazl employed the older solar chronologies. For example, when giving the date of Akbar's birth he used the Jalali and Yazdegird eras in addition to the Hijra.[88]

As he began to tell the story of Akbar's reign, Abu al-Fazl faced a chronological dilemma. Akbar had come to the throne in 1556 but the *Tarikh-i Ilahi* was not introduced until 1584. To solve this problem the historian inserted two chapters (3 and 4) on the new era at the very beginning of his history, just after the initial chapters on the young emperor's accession. Thereafter the primary chronological system was the *Tarikh-i Ilahi* and the significant dates were solar. However, when a new year was announced or an Islamic festival was described, the Hijra year equivalent was often given. For example, in dating Id-i Qurban of 1556 the historian wrote "... on the day of Bahman, 2 Aban Divine Month, corresponding to Thursday, 10 Zu al-Hijja, which was the Id-i Qurban ..."[89] And his heading for chapter 7 of vol. 3 read "Beginning of the 18th Divine Year from the Accession, viz., the Year Shahriyar of the Second Cycle." But in the first paragraph he gave the Hijra era equivalent.[90]

Akbar's son, Jahangir, followed his father's lead in chronology as in other matters. His memoir – *Tuzuk-i Jahangiri* – was divided according to *Tarikh-i Ilahi* regnal years. For example, the chapter heading for the fifteenth year read: "The Fifteenth New Year's Feast after the Auspicious Accession."[91] The first sentence gave the Hijra era date of Nau Ruz but the second began. "On Saturday, the 2d (Farvardin) ..."[92] In the *Tarikh-i Ilahi* calendar Nau Ruz (March 21) was 1 Farvardin. In the same way Jahangir would often mix the two eras, giving a complete Hijra Era date (day, month, year), followed several sentences later by a solar era day and month.[93]

By the reign of Akbar's grandson, Shahjahan, opposition to the new era had begun to increase. In 1631–1632 Mulla Farid al-Din Ibrahim Dihlavi, a student of Shirazi and Shahjahan's chief astrologer,

[87] Rizvi, *Religious and Intellectual*, 262–77.
[88] Abu al-Fazl, *Akbar Nama*, 1: 54.
[89] Ibid., 2: 51.
[90] Ibid., 3:1, 45.
[91] Jahangir, *Tuzuk*, 2: 130.
[92] Ibid.
[93] See, for example, 1: 294–5.

drafted a revised (but never implemented) version of the *Tarikh-i Ilahi* – the *Tarikh-i Ilahi-i Shahjahani*.[94] And in 1638 Mirza Amin Qazvini completed his history of the ruler's first decade (the *Padshahnama*). Employing the *Tarikh-i Ilahi*, he began each regnal year on Nau Ruz (1 Farvardin). However, when the court historian presented his account to the emperor, it was not approved. A format which had been *de rigueur* for the previous forty years had suddenly become unacceptable.[95] Influenced no doubt by the conservative ulama of the day, Shahjahan declared the solar era heterodox and appointed a new historian, Abd al-Hamid Lahori. Lahori revised Qazvini's draft, replacing the Divine Era chronology with the Hijra. Because the new division necessitated a reshuffling of the narrative – moving the important incidents of Shahjahan's first decade from one year to the next – it was much more than a simple relabeling of chapters. Despite the official censure, Lahori did not entirely expunge Akbar's new era from his history. He continued to supply both the solar and lunar dates for Shahjahan's birthday and to date and describe the Nau Ruz festivals, which continued to be lavishly celebrated, even after they no longer marked either the first day of the year or the anniversary of the emperor's accession.[96]

Aurangzeb, the last of the Great Mughals and the most traditional of Akbar's descendants, reacted most strongly to the new era. In 1659, the second year of his reign, he abolished the celebration of Nau Ruz, and in 1668, at the end of his first decade, he discontinued the celebration of his lunar birthday. While Muhammad Kazim, his official chronicler, was allowed to compile an account of his first decade (employing the Hijra Era exclusively), Aurangzeb decided, after receiving the *Alamgir Nama*, to abandon the enterprise altogether. Recording and glorifying a ruler's achievements was antithetical to the pious life, and from that point on the emperor forbade the keeping of official records and the writing of official histories.[97]

Khafi Khan (1663–ca. 1731), writing some thirty years after Aurangzeb's death, described the practical drawbacks of the emperor's decision:

94 H. J. Winter, "Persian Science in Safavid Times," in Jackson, ed., *Cambridge History of Iran: 6*, 593.

95 Khan, *Shahjahan Nama*, xxiii–xxv.

96 Ibid., 21.

97 Khafi Khan, "*Muntakhab al-Lubab*," in Dowson, ed., *History of India*, 7: 174.

> Mathematicians, astronomers and men who have studied history, know that . . . the recurrence of the four seasons, summer, winter, the rainy season of the Hindus, the autumn and spring harvests, the ripening of the corn and fruit of each season, the *tankwah* (income) of the *jagirs*, and the money of the *mansabdars*, are all dependent upon the solar reckoning, and cannot be regulated by the lunar.[98]

OTTOMAN EMPIRE

Fiscal

For the Ottoman empire, given its location and self-image, it is no surprise to find that the indigenous solar chronological tradition was Christian. The calendar, called the Rumi (Roman) or Maliye (Fiscal), was the Julian with a New Year's Day of 21 March (Nau Ruz). The Ottoman solar era, however, was not the Christian Anno Domini. Called the *Sene-i Maliye* (Fiscal Era) or *Sene-i Rumiya* (Roman Era), it began as way of keeping track of agricultural taxes and expenditures in a primarily agrarian economy. The Ottoman solar era, however, unlike those of the Mughals and the Safavids, was not completely independent of the Hijra. Rather, like the Kharaji Taxation Era under the Umayyads or the Fasli Taxation Era under Akbar, its number sequence was tied to the Hijra and had to be periodically adjusted. For the Ottomans this adjustment was called *sivis* (skipping). Like *izdilaq* (sliding) for the Umayyads and *tahvil* (changing) for the Mughals, it involved periodically eliminating one year in the solar era numbering sequence.

First employed during the reign of Bayezid I (1481–1512),[99] the Rumi Era was initially a bookkeeping device, found only in taxation and salary documents and records. For example, in an almanac (*takvim*) drawn up during the reign of Bayezid's son, Selim I (1512–1520), there were columns for the Yazdegird, Jalali, and Turkish Twelve-Year Animal eras but none for the Rumi.[100] Over the years, however, as the economic and political problems associated with two eras became more serious, Ottoman officials began to tinker with the Rumi Era, establishing rules and procedures that made it increasingly accurate and public.

Because the Ottomans paid the Janissaries (infantry) of the imperial household in cash and because their salaries were traditionally due at the

98 Ibid., 7: 241–2.

99 Darling, *Revenue-Raising*, 135.

100 Massumeh Farhad and Serpil Bagci, "The Manuscripts," in Bagci and Farhad, eds., *Falnama*, 50–1.

end of each lunar quarter, the chronological mismatch between income and expenditures became increasingly difficult to manage. In the Janissary payroll schedule the four Hijra quarters were given names. Constructed from the first letters of the three months, they were: Masar (Muharram, Safar, Rabi I), Reced (Rabi II, Cuma I, Cuma II), Reshen (Rajab, Shaban, Ramazan), and Lezez (Sheval, Zu al-Qada, Zu al-Hijja). Agricultural taxes, by contrast, were collected on Nau Ruz (Nevruz, 11 or 21 March) and 1 August.[101]

The difficulty then was the eleven-day difference between the lunar Hijra era year (354 days) and the solar Rumi era year (365 days). Because the salary year was shorter than the revenue year, expenditures would eventually outrun income. After the revenue of eight solar years had been collected, salaries for eight and a quarter (8 x 11= 88 days) lunar years would have been paid. And after thirty-two solar years, an extra year of lunar salaries would come due.

To solve this problem the Ottomans employed several strategies: raising taxes, devaluing the currency, borrowing from the Inner Treasury, obtaining extra revenue from military conquests, or saving. None of these solutions, however, was foolproof, and over time the crises that arose in the thirty-third lunar year became increasingly burdensome. In an intriguing article Prof. Halil Sahillioglu analyzed the Ottoman efforts to deal with the "*sivis*" year deficits.[102]

In the late fourteenth and early fifteenth centuries the problem was less serious. The Janissaries were only a small percentage of Ottoman forces (about 5 percent in the last years of Murad II [1421–1451]), and the empire was rapidly expanding. However, from the late fifteenth century through the early eighteenth the "*sivis*" deficits became harder to fund. In 1448 Mehmed II decided to devalue the currency to meet the accumulated salary arrears. The conquest of Istanbul three years later, however, alleviated most of his financial pressures.[103] In 1481, during the last year of his life, the aging sultan had to devalue once again and, as before, was met with a rebellion by the Janissaries. In 1513 Selim I put off the immediate fiscal pressure and was saved by his conquest in 1517 of Egypt and Syria.[104] In 1546 the deficit was covered by surplus revenues from the new provinces

101 H. Sahillioglu, "Sivis Year Crises in the Ottoman Empire," in M. A. Cook, ed., *Studies in the Economic History of the Middle East* (Oxford: Oxford University Press, 1970), 230–54.

102 Ibid.

103 Ibid., 238.

104 Ibid., 239.

and by new taxes. Six years later, however, devaluation was needed.[105] In 1578 the wars with the Safavids, unlike the conquests of Istanbul and Egypt/Syria, drained rather than replenished the imperial coffers. In 1581 additional taxes were levied, and in 1584 the usual devaluation was ordered. In 1586 the Janissaries rose in anger – officials of both the treasury and the mint were killed in the melee.[106] In 1612 the deficit was covered from the balances of the Inner Treasury. Under the Ottomans the Outer Treasury contained the ordinary revenues of the state and was managed by the Grand Wazir. It underwrote the everyday expenses of administration and security. The Inner Treasury, by contrast, was the personal preserve of the sultan. Its sources were the Baghdad customs, the confiscated estates of high officials, and the surpluses of the Outer Treasury.

In 1644 the murder of the Grand Wazir and the extravagant spending of Sultan Ibrahim (1640–1648) exacerbated the usual deficit. Additional taxes and devaluation solved the immediate problem but led to another Janissary revolt.[107] In 1677 a transfer from the Inner Treasury covered the shortfall. At this time the Head Treasurer formally adopted the *Maliye Financial Era* (renaming it the *Rumi Takvim* or *Roman Era*), raising it from an informal, interdepartmental method of managing income and expenditures to an official imperial chronology.[108] Although in 1710 the Grand Wazir had to resort to short-term borrowing, he also decided to again reform the accounting system; thereafter the treasury calculated the annual budget using both the Hijra and Roman Eras.[109] In 1789 Selim III (1789–1807) instituted the *Sana-i Maaliyya* (*Financial Era*), changing the names of the nine Julian months to their Syrian counterparts.[110] And in 1840 during the Tanzimat period of modernization and reform the Roman Era became the official dating system for all civic, nonreligious activities. The calendar year began on 1 March and the months were renamed but the years continued to be tied to the Hijra: "skipping" was not eliminated. As a result, in 1871 the "*sivis*" adjustment was forgotten and for the rest of its history the Roman Era was one year behind the Hijra. The discrepancy remained until the adoption of the Gregorian Era by Ataturk in 1926.[111]

105 Ibid., 240.
106 Ibid., 240.
107 Ibid., 241–2.
108 Ibid., 242–3.
109 Ibid., 247–8.
110 Taqizadeh, "Various Eras, 915.
111 *Encyclopaedia of Islam*, 2d. ed., s.v. "Tarikh," 263.

Dating

For Ottoman historians and record keepers of the early modern period the problem of chronology was relatively uncomplicated. Unlike their counterparts in the Mughal and Safavid empires, they did not have to introduce solar era dates into their narratives nor did they have to calculate equivalent dates across chronologies. In any public writing – orders, proclamations, inscriptions, histories, or reports – the Hijra Era was the sole system of chronology.

The earliest Ottoman historical accounts were compiled from *takvims* (almanacs). These were chronological lists of important events – both political and natural.[112] In these early *takvims* there were no dates. Rather significant events were situated in relationship to the year of the almanac. For example, "It is four years since the castle and city of X was conquered." With the accession of Murad II (1421–1451), however, there came an abrupt change of style; the events of his reign were recorded in a series of short paragraphs, very similar to those in earlier imperial almanacs, but each year's entry concluded with a Hijra date.[113] Two early examples survive – one from 1444–1445 and the other from 1447–1448. These were produced for the sultan by the *muneccimbashi*. Based on the recently completed *Zij i-Sultani* of Ulugh Beg, each opened with a brief historical section: the Islamic prophets, the early caliphs, and the rulers and major events of the Seljuq, Karaman, and Ottoman dynasties. An overall prediction followed along with astronomical and astrological information for each day of the year.

During the late fifteenth and early sixteenth centuries Ottoman historians utilized *takvims, menakib names* (semilegendary accounts of famous men), and the *ghaza names* (accounts of military exploits) to produce relatively straightforward narratives – like the *Tevarih-i Al-i Osman* (*Chronicles of the House of Osman*).[114] By the mid-sixteenth century, however, the historiographical tradition had shifted. The sultans began to appoint official historians called *shehnameci* (writers of *shehnames*). The title, of course, referred to the national epic of Iran, the *Shahnameh* (*Book of Kings*), written by the Persian poet, Firdausi (ca. 935–1020).

112 Halil Inalcik, "The Rise of Ottoman Historiography," in Bernard Lewis and P. M. Holt, eds., *Historians of the Middle East* (Oxford: Oxford University Press, 1962), 152–67.

113 V. L. Menage, "The Beginnings of Ottoman Historiography," in Ibid., 171–2.

114 Cornell Fleischer, *Bureaucrat and Intellectual in the Ottoman Empire: The Historian Mustafa Ali (1541–1600)* (Princeton, NJ: Princeton University Press, 1986), 237–40; *Encyclopaedia of Islam*, 2d. ed., s.v. "Tarikh," 290–96.

That the Ottomans at the height of their wealth and power should have chosen this title for their official historian testifies to the prestige of Persian poetry and prose at the imperial court. Five men occupied the position: Suleiman I appointed the first, Fethulah Afif Celebi, in 1545, and when Ahmed I (1603–1617) did not appoint a successor, the post came to an end. The works produced by these men varied in language (Persian or Turkish) and in style (prose or poetry). They included world histories, accounts of individual sultans, and descriptions of military campaigns. Set in the mythological world of the Persian epic, these histories functioned partly as imperial propaganda: the sultan as universal ruler, mystic guide, or even (in the case of Suleiman) shadow of god.[115] After Ahmed I the office of official historian remained vacant for over one hundred years, until the appointment of Mustafa Naima (1656–1716) in 1709.

In the interval between the two official historians, however, the activities of the sultans did not go unrecorded. For example, the administrator and man of letters Mustafa Ali, although he had no title, composed the *Essence of History* (ca. 1592–1598).[116] The work had four parts:

1. creation of the world and of man
2. pre-Islamic prophets and Islamic history down to the Mongols
3. Mongol and Turkic dynasties
4. Ottomans.

The last part was organized by ruler, each reign divided into a number of named and dated "events."[117] For Ali, as for the other Ottoman historians, the chronological framework was the lunar Hijra.

The *Tarih-i Naima* of Mustafa Naima covered Ottoman history from the beginning of the millennium (1000 AH/1591 CE) to 1070 AH/1660 CE. A typical Muslim chronicle, it treated each year as a unit and each event in that year as a single entry. An important occurrence, however, such as a major campaign, might extend over several entries and, if prolonged over several years could only be followed in its entirety by skipping from year to year and entry to entry.[118] Naima's chronology was typically Ottoman: Only Hijra Era days, months, and years. The New Year began on 1 Muharram and no mention of Nau Ruz (21 March) was found.[119]

115 F. Sinem Eryilmaz, "The Official Ottoman Shehname: An Instance of Cultural Memesis," paper presented at the Sawyer Seminar, University of Chicago, 2001–2002.
116 Fleischer, *Bureaucrat and Intellectual*, 235–7.
117 Ibid., 245–52.
118 Lewis Thomas, *A Study of Naima* (New York: New York University Press, 1972), 126–7.
119 Ibid., 100.

Because, however, of their exclusive use of the liturgical era, the millennium was an important organizing principle for Ottoman historians. Ali opens the Ottoman section of his history by listing the other principal Muslim states – Safavid, Mughal, and Uzbek – in 1000 AH.[120] And Naima, beginning his history at the millennium, included an uncharacteristic (for an Ottoman historian) discussion of eras:

> The thousandth year of the Hijrah commenced on the first of Moharrem – which is also the Sabbath day (Saturday) – and is the three hundred and fifty-four thousandth four hundred and ninety-ninth day of the prophet's flight. ... This year of happy omen, connected with antecedent time, makes the 184,693rd year [since the beginning of the world], and the 5350th of the Jewish era; but, according to Melek'l mowid's mode of reckoning, [it] is the 7216th year. The thousandth year of the Hijrah is the 4974th from Noah's flood, but which, according to the calculations of astronomers, falls short of that period by a space of 270 years: and the 2807th of the supreme conjunction [Grand Conjunction of Jupiter and Saturn]; the 2338th of the era of Nebuchadnossar; the 1901st of the era of Alexander; the 1590th of the Christian era; the 1360th of the Copts; the 959th of the Yazdijerd [the ancient Persian era]; the 533rd of Jellali [the modern Persian era]; and the 167th of the middle or intermediate supreme conjunction.[121]

So central was the millennium to Ottoman histories that one scholar has argued that most chronicles that included the eleventh Hijra century (sixteenth and seventeenth centuries CE) had as their theme "the history of the year 1000 (Tarih-i Sene-i Elf)."[122]

For the early modern Islamic empires the Hijra Era presented a problem. Divinely revealed, it organized the ceremonial occasions of the liturgical year while recording the defining events – dynastic, military, architectural, and religious – of the community's past. As the chronological framework for an agrarian economy, however, it was seriously deficient. As a result, from the seventh century CE onward rulers and administrators began to introduce new temporal systems, solar eras that would enable tax collectors and record keepers to more accurately manage revenues and expenditures.

These first solar eras, created to solve the fiscal problem, brought their own sets of difficulties. For the Kharaji Taxation Era of the Umayyads and Abbasids, as for the Fasli and Rumi Eras of the Mughals and Ottomans, the principal shortcoming was the institutional memory required – every thirty-three years the numbering sequence of the new era had to be

[120] Fleischer, *Bureaucrat and Intellectual*, 54.

[121] *Annals of the Turkish Empire, from 1591 to 1659 of the Christian Era by Naima*, trans. Charles Fraser; reprint ed. (New York: Arno Press, 1973), ix.

[122] Thomas, *Study of Naima*, 135.

adjusted. Although the early modern empires were larger, wealthier, and better organized than their medieval predecessors, none was able to completely escape the dilemmas raised by a lunar era in an agrarian state.

Because the Safavids had the smallest population, were less centralized and monetized, and had a long history of solar chronologies, they were more easily able to deal with the fiscal problem. The three most common solar eras – the Turkish Twelve-Year Animal, the Yazdegird, and the Jalali – were all of Iranian origin, and all three featured the Zoroastrian calendar, New Year's Day falling on the Vernal Equinox. As a result, scheduling the collection of taxes and the payment of salaries was relatively straightforward.

For the Mughals and Ottomans, by contrast, fiscal solutions were more complicated. Like the Safavids, the Mughals had inherited a rich chronological tradition. On the Indic side there were multiple calendars and eras – both solar and lunisolar – and numerous astrologers skilled in the preparation of almanacs and horoscopes and the translation of dates from one era and calendar to the other. On the Islamic side, the Timurids had bequeathed the Turkish Animal Era and Ulugh Beg's *Zij-i Sultani* while the Persians had contributed the Zoroastrian calendar and the Yazdegird and Jalali eras. Unlike the Safavids, however, the Mughals ruled a populous, rich, and culturally complex empire. The annual monsoon and the two great river systems supported an agrarian economy of two harvests a year and 125–150 million people – ten times that of the Safavids and triple that of the Ottomans. Although Akbar's introduction of the *Tarikh-i Ilahi* in 1584 may have been tangentially related to the fact that the "change" or "skip" year of the Fasli Era was fast approaching (in 1589), the new era is probably better understood as part of Akbar's policy of "lasting reconciliation" – a new chronological system for a new imperial order. In addition to the symbolic and fiscal reasons for its introduction, the new era was also intended to address a commercial problem. Although the economy of the Mughal Empire was overwhelmingly agrarian, a large portion of the domestic economy was monetized. Overseas trade in cloth, spices, indigo, and sugar had flourished for a least a millennium. Thus, an additional rationale for the new era was the simplified, abbreviated notation it would bring to the accounting and recordkeeping practices of merchants and bankers.

The Ottomans faced a different kind of problem. Unlike the Umayyads and Abbasids, they had the administrative expertise to properly adjust the "skip" years. And, like Afushtah Natanzi, the Safavid historian, they expressed their solar years in lunar terms – one 365-day Rumi year equaling twelve lunar months plus eleven days. But the Ottomans, while they were able to translate the dates of agrarian tax collections into their lunar equivalents, could never establish a revised salary schedule for the Janissaries. The payment of the imperial infantry four times a year (at the beginning of each Hijra quarter) was a tradition too long established to be tampered with. Thus, despite the fiscal and political crises that periodically erupted, the Ottomans never attempted to introduce a solar era that was independent of the Hijra. And perhaps, given their self-image as successors to the divinely guided caliphs and protectors of the holy cities, they were right not to try. The fate of Akbar's new era – successively diluted and eventually abandoned – suggests that the hold of the traditional system of religious chronology on the local populace was extremely hard to break.

The various eras caused difficulties for historians as well as for tax collectors and administrators. How were they to date their narratives? What chronological system or systems should they employ?

For Safavid historians, the chronological problem was especially complicated. The long history of solar eras in Iran meant that a simple Hijra system would not suffice. Given the array of choices, it is not surprising to find the four Safavid historians adopting different strategies. It is, however, ironic that Iskandar Munshi, so conscious of the chronological pitfalls, should have made so many errors. He consistently failed to provide the proper Hijra equivalents for the animal and regnal years and for Nau Ruz. The other three historians, however, each taking a different approach, managed to avoid most of Iskandar Munshi's mistakes. The most accurate was Abbas's chief astrologer, Jalal al-Din Munajjim Bashi. Himself a skilled chronologer, he had access to the best almanacs and the best advice – able to draw on the libraries and talents of the astrologers of the imperial household. While both Afushtah Natanzi and Mirza Beg Junabadi had different strategies, their choices were driven by the fact that their histories covered much less ground than the first two. Natanzi divided his chapters according to regnal year, using Hijra era dates. His year, however, was the 365-day solar rather than the 354-day lunar, and his New Year and accession anniversary both fell on the Vernal Equinox. His Hijra equivalents for Nau Ruz, however, were accurate. Mirza Beg Junabadi's approach was the most reductive. Because his history was limited to the Safavids, his account of Shah Abbas was quite long. Most of the chapters, however, were not divided according to regnal year but rather according to incident – a battle, a marriage, a festival, the founding of a city, or the building of a mosque. And, often, his chapters included no date whatsoever or, if they did, the Hijra year only.

Mughal historians also had a variety of eras to choose from. For an early modern ruler Akbar took an unusual interest in chronology. In the first official history of his reign (the Millennium History) he introduced a new era, and after 1584 the *Tarikh-i Ilahi* became the chronological standard for all public writing. Nizam al-Din Ahmad's chapters were organized according to Divine Era regnal years (though the dates of individual events were more often Hijra than Ilahi) while Abu al-Fazl in the *Akbar Nama* relied almost exclusively on the new era. Although Nizam al-Din committed a few dating errors, the chief problem for Mughal historians was not calculating chronological equivalents but was rather countering the rising opposition to the use of any solar era whatsoever. These misgivings, first seen in Badauni's work, had, by the reign of Akbar's grandson, became increasingly common. The official history of Shahjahan's first decade had to be withdrawn, and, after receiving his Hijra-only history, Aurangzeb, Shahjahan's successor, decided to give up official history writing altogether.

Ottoman historians, by contrast, did not face the chronological confusion of their Safavid and Mughal counterparts. The Rumi Era remained tied to the Hijra and never became a widely employed public system until the nineteenth century. As a result, early modern Ottoman historians never had to worry about calculating equivalencies across eras or defending the use of a nontraditional chronological system. Nevertheless, in the late sixteenth and early seventeenth centuries their complete reliance on the Hijra Era did give their narratives a characteristic millenarian focus. In contrast, the end of the first Islamic millennium (1591) was almost completely ignored by the Mughal and Safavid historians. Although Akbar's commissioning of the Millennium History (1582) would seem to indicate a certain awareness, the imposition of a different era for that chronicle and the introduction of the *Tarikh-i Ilahi* (1584) suggest that commemorating one thousand years of the religion was of no particular importance. Although we would expect Abu al-Fazl to have disregarded the event,[123] it is strange to find Abd al-Qadir Badauni doing the same thing. Having declared the *Tarikh-i Ilahi* heretical, he divided his chronicle into Hijra-year chapters. Nevertheless, the millennium year segment began with the Hijra date of Nau Ruz, the regnal year, and nothing more.[124] Safavid historians took a similar approach. Iskandar Munshi began his account of Abbas's fifth year (1591–1592) with the animal year, the regnal year, and the Hijra date of Nau Ruz.[125] The end of the first Islamic millennium was passed over entirely.

[123] Abu al-Fazl, *Akbar Nama*, 3: 927.
[124] Badaoni, *Muntakhab*, 2: 392.
[125] Munshi, *Tarikh-e Alamara*, 2: 613.

5

Chronology: Millenarian

In the early modern Islamic world, the late sixteenth century was a chronological watershed. Two millenarian events marked the period: On the one hand, a Grand Conjunction of the planets Jupiter and Saturn occurred in 1583 while, on the other, the first millennium of the Hijra Era ended in 1591. This rare and unusual combination created an atmosphere of apocalyptic expectation, each empire giving birth to a unique mix of prophets, movements, and ideologies.

Although a developed theory of conjunction or historical astrology was first found in the works of two eighth-/ninth-century Iranian *munajjims* – Mashaallah ibn Athari (ca. 740–815) and Abu Mashar al-Balkhi (787–886) – its origins can be traced to the first half of the first millennium. In the Sassanid Pahlavi treatise – *Zik-i Shatro-Ayar* (Persian, *Zij-i Shahi* or *Zij-i Shahriyaran al-Shahi*) – the conjunctions of Jupiter and Saturn were analyzed in the context of a more comprehensive chronological theory – the world year. Beginning with a grand conjunction of all planets at the zero point of the ecliptic (the first point of Aries or the Vernal Equinox), a world year lasted for 360,000 years (according to the Babylonians and Iranians) or 4,300,000 years (according to the Indians). Each year so defined ended with a great cataclysm – the universe was completely destroyed and then re-created. The first cosmic destruction had been by water and the second would be by fire. The Jupiter–Saturn conjunction cycles, on the other hand, were much shorter, and their effects less cataclysmic.[1]

[1] E. S. Kennedy, "The World Year Concept in Islamic Astrology," in Kennedy, *Studies in the Islamic Exact Sciences*, 351–71; E. S. Kennedy and B. L. van der Weerden," The World Year of the Persians," in Ibid., 338–50; E. S. Kennedy, "The Sasanian Astronomical

Mashaallah ibn Athari was a Jewish convert from Basra. An Abbasid *munajjim* who knew the *Zij-i Shahi* and the Greek and Indian treatises (through Pahlavi and Arabic translations), he introduced the latest theories to the early Arab astronomers/astrologers. He was part of the group that determined the proper time (30 July 762) to found the Abbasid capital of Baghdad, and he authored works on all aspects of astrology. Of his twenty-eight treatises, the most influential were: *Kitab al-Mawalid* (*Great Book of Nativities*), *Fi al-Qiranat wa al-Adyan wa al-Milal* (*On Conjunctions, Peoples, and Religions*), *Kitab Sanat al-Asturlab wa al-Amal Biha* (*Book of the Construction of an Astrolabe and its Use*), *Kitab al-Duwal wa al-Milal* (*Book of Dynasties and Religions*), and *Ahkam al Qiranat wa al-Mumazajat* (*Judgments of Conjunctions and Mixtures*).[2] Known in medieval and early modern Europe as Messala or Messahalla, his "On Conjunctions, Peoples, and Religions" was extraordinarily popular, serving, for example, as the source of one of Chaucer's horoscopes.[3]

It was Abu Mashar, however, who became the most prolific and influential astrologer of the premodern Eurasian world. A staple of the madrasa curriculum, he influenced orthodox and heterodox thinkers alike – from the eleventh century Ismailis and the fifteenth- and sixteenth-century Nuqtavis to the everyday practitioners of the early modern imperial courts. His authority, however, was not limited to the Islamic world, and from the twelfth century onward his writings (translated into Latin) swayed astrologers, philosophers, and statesmen in the centers of European science and politics.

The eastern Iranian city of Balkh, Abu Mashar's birthplace, was a hub of cultural and religious diversity – filled with Indians, Chinese, Scythians, Greco Syrians, and Iranians who were Jews, Nestorians, Buddhists, Hindus, and Zoroastrians. He was a third-generation member of the Iranian, Pahlavi-oriented elite who had played such an important role in the recently established Abbasid empire (750–1258). Despite his eclectic background, however, he was a dedicated Muslim, beginning his career in Baghdad as a hadith expert. A quarrel with the great Arab philosopher

Handbook Zij –I Shah and the Astrological Doctrine of the Transit (Mamarr)," in Ibid., 319–35; Kathryn Babayan, *Mystics, Monarchs, and Messiahs: Cultural Landscapes of Early Modern Iran* (Cambridge, MA: Harvard University Press, 2003), 13–4, 21–3, 36–7; *Encyclopaedia of Islam*, 2d ed., s.v. "Kiran."

2 *Dictionary of Scientific Biography*, s.v. "Masha'alah."

3 E. S. Kennedy, "A Horoscope of Messehalla in the Chaucer Equatrorium Ms.," in E. S. Kennedy, *Studies in the Islamic Exact Sciences*, 629–30; E. S. Kennedy and David Pingree, *The Astrological History of Masha`allah* (Cambridge, MA: Harvard University Press, 1971) v–vii, 69, 75.

al-Kindi (ca. 796–873) turned him toward "mathematics" (arithmetic, geometry, music, astronomy, and astrology). From then on his energies were devoted to a philosophical justification of astrology, and he constructed a system that reflected the three intellectual strands of his heritage: the Pahlavi Greco-Indo-Iranian tradition of the *Zij-i Shahi*, the Sanskrit Greco-Indian tradition of the *Zij-al Sindhind*, and the Greek tradition of Aristotle and Ptolemy. In Abu Mashar's *Zij-i Hazarat* (*Treatise of the Millennia*), for example, his computation of the mean motions of the planets employed the *yuga* technique of the *Zij-al Sindhind*, his prime meridian and planetary parameters were taken from the *Zij i-Shahi*, and his overall planetary model was Ptolemaic.[4]

Of the forty-two works of Abu Mashar, the most important were: *Kitab al-Madkhal al-Kabir 'ala il Akham al-Nujum* (*Great Introduction to the Science of Astrology*), *Zij–i Hazarat* (*Treatise of the Millennia*), *Kitab al-Mawalid al-Kabir* (*Great Book of Nativities*), *Kitab al-Qiranat* (*Book of Conjunctions*) or *Kitab al-Milal wa al-Duwal* (*Book of Religions and Dynasties*), *Kitab Tahawil Sini al-Alam* (*Book of the Revolutions of the World Years*), *Kitab al-Uluf* (*Book of the Thousands*), and *Kitab Zij al-Qiranat wa al-Ikhtaraqat* (*Tables of Conjunctions and Transits*).

Although Abu Mashar's works spanned the entire field of medieval astrology, his (and Mashaallah's) principal contribution to the Islamic concept of time was the chronological system based on the conjunctions of Jupiter and Saturn. The two largest and most distant planets, Jupiter was about five hundred million miles from the sun and Saturn about nine hundred million (the Earth was ninety-three million.) As a result, it took Jupiter about twelve years to complete one revolution around the sun and Saturn about thirty. Every twenty years or so, when the two planets appeared more or less together, a conjunction occurred. In Abu Mashar's theory, though, all conjunctions were not equal. Tracked across the sky through the twelve signs of the zodiac, the Jupiter-Saturn juxtapositions that marked the shifts from one triplicity (i.e., group of three signs) to the next and occurred every 240 years were the most important.[5] There were four

[4] *Dictionary of Scientific Biography*, s.v. "Abu Mashar;" David Pingree, *The Thousands of Abu Ma'shar* (London. The Warburg Institute, 1968); *Abu Ma'sar on Historical Astrology*, ed. and trans. Keiji Yamamoto and Charles Burnett, 2 vols. (Leiden: Brill, 2000); E. S. Kennedy, "The World year Concept in Islamic Astrology," 351–71; E. S. Kennedy and B. L. van der Weerden," The World Year of the Persians," 338–50.

[5] E. S. Kennedy, "An Astrological History Based on the Career of Genghis Khan," in E. S. Kennedy, *Astronomy and Astrology in the Medieval Islamic World* (London: Ashgate Variorum, 1998), 224.

triplicities, each named after one of the four Aristotelian elements: the first, Fire, included the zodiacal houses of Aries, Leo, and Sagittarius; the second, Earth, contained Taurus, Capricorn, and Virgo; the third, Air, had Gemini, Libra, and Aquarius; and the last, Water, included Cancer, Scorpio, and Pisces. The most significant conjunction of all, however, was the Grand Conjunction. Occurring once every 960 years, it marked the completion of a full revolution – when the Jupiter–Saturn conjunctions had cycled through all twelve signs of the zodiac and had returned to the first point of Aries, shifting from the Watery to the Fiery triplicity.[6]

The *Kitab al-Qiranat* contained the most complete description of Abu Mashar's system. In this treatise the author demonstrated that the great (and not so great) events of the past had been marked by conjunctions of greater (or lesser) rarity. For example, the Grand Conjunction of 3101 BCE (17–18 February) indicated both the biblical Great Deluge and the beginning of the Indic Kali Yuga. Other Jupiter–Saturn conjunctions were equally momentous, coinciding with the birth of the prophet Muhammad (571 CE) and the founding of the Abbasid Empire (749). Although neither Abu Mashar nor Mashaallah had much to say about the future, the calculations of later *munajjims* pinpointed an early modern heavenly event of enormous importance. The year 1583 (sometimes miscalculated as 1582) would include another Grand Conjunction, a once in a millennium happening.[7] For Muslims this conjunction would be especially significant since it would occur 960 years (a full conjunction cycle) after the founding of the religion (622).

In the first two centuries after his death Abu Mashar's theories were widely read. Of particular importance was his impact on the Ismailis, a Shiite sect that spread rapidly during the rule of the Fatimid Caliphate (969–1171). The Ismaili Shiites of the tenth and eleventh centuries shared many beliefs and practices with the Imami Shiites of Safavid Iran. Both traced their disagreements with Sunni Islam to the leadership struggle that erupted after the death of the prophet. Although both sects maintained that the descendants of Ali, the prophet's son-in-law and the fourth caliph, were the divinely sanctioned leaders of the community, they differed on the exact order of succession. The Ismailis (or Seveners) maintained that Ismail ibn Jafar was the seventh Imam, the true successor to Jafar al-Sadiq, the sixth

[6] Ibid., 223–31.

[7] See Margaret Aston, "The Fiery Trigon Conjunction: An Elizabethan Astrological Prediction," *ISIS* (1970): 159–87.

Imam. The Imamis (or Twelvers), on the other hand, believed that Musa al-Kazim, Ismail's younger brother, had been the true Imam, and they traced their line through him to the twelfth Imam. Although in the eyes of the Sunni majority Seveners and Twelvers were equally heretical, the Ismaili belief system was more esoteric, unusually open to unorthodox ideas and practices. Imami philosophy and theology, on the other hand, was more literal and orthodox, stressing the importance of sharia (Islamic law) and hadith (traditions of the prophet).

In the transmission of Abu Mashar's conjunction astrology from the eighth century to the sixteenth the works of two Ismaili authors were of critical importance. The first was the author of the *Rasail Ikhwan al-Safa (Epistles of the Brethern of Purity)*. A mysterious secret society that most scholars think was Ismaili, the Ikhwan al-Safa were based in tenth-century Basra (the birthplace of Mashaallah). The fifty-two chapters of the *Rasail* were divided into four sections: mathematics, natural sciences, psychology, and theology. In the chapter on astronomy Abu Mashar was mentioned by name, and his system was briefly outlined. His determination that the conjunction of 571 had marked the prophet's birth was included as well as his explanation of the Sassanid world year: a 360,000-year epoch bookmarked by two Grand Conjunctions – the first in 183,101 BCE and the second in 176,899 CE.[8]

The second Ismaili author was the eleventh-century poet, traveler, and religious leader Nasir-i Khusrau (1004–1088). Born in Balkh (Abu Mashar's birthplace), he was another Islamic polymath (like the Mughal Mir Fathullah Shirazi and the Ottoman Taqi al-Din). Having mastered mathematics, the natural sciences, medicine, and Greek philosophy, Nasir-i Khusrau spent the first forty years of his life as a financial secretary and revenue collector. In 1046, however, he threw it all over and embarked on an epic six-year journey through the eastern Islamic world, visiting Mecca four times and residing in Cairo, the capital of the Fatimid empire, for three years. In Cairo he read Ismaili theology and philosophy, including the *Rasail Ikhwan al-Safa*, and became a passionate convert. His talent and dynamism brought him to the attention of the Fatimid Caliph al-Mustansir (1036–1094), who sent him back to his native Khurasan as a missionary. Only moderately successful as a preacher, his fame rested on his writing.

[8] Farhad Daftary, *A Short History of the Ismailis* (Edinburgh: Edinburgh University Press, 1998); Abbas Hamdani, "A Critique of Paul Casanova's Dating of the Rasa'il Khwan al-Safa," in Farhad Daftary, ed., *Medieval Ismaili History and Thought* (Cambridge: Cambridge University Press, 1996), 145–52.

Although the *Safarnama* (*Book of Travels*) is his best-known composition, he wrote several other works: a diwan of poetry, a philosophical treatise, a book on mathematics, and two long narrative poems – the *Raushana-i Nama* (*The Book of Enlightenment*) and *the Saadatnama* (*Book of Felicity*). *The Raushana-i Nama* contains excerpts from the *Rasail*, and his *divan* (collection of poetry) contains several poems with conjunction themes.[9]

The second millenarian event of the late sixteenth century was the end in 1591 (1 Muharram 1000/19 October 1591) of the first one thousand years of Islamic history. A symbolic occasion of great moment, it sparked the reappearance across the early modern Islamic world of two venerable apocalyptic figures: the Mahdi (Guided One) and the Mujaddid (Renewer).

The word Mahdi does not appear in the Quran or in the earliest hadith collection. It is, however, found in the five later canonical compilations. According to one early hadith: "The Messenger of God said: "The earth will be filled with injustice and crime ... God will send a man from me.... He will fill [the earth] with equity and justice ..."[10] And, according to the historian Ibn Khaldun (1332–1382):

> It has been well known ... by Muslims in every epoch, that at the end of time a man from the family (of the prophet) will without fail make his appearance, one that will strengthen the religion and make justice triumph. The Muslims will follow him, and he will gain domination over the Muslim realm. He will be called the Mahdi. Following him, the anti-Christ will appear, together with all the subsequent signs of the Hour (the Day of Judgment). ... After (the Mahdi), Isa (Jesus) will descend and kill the anti-Christ. Or, Jesus will descend together with the Mahdi, and help him kill (the anti-Christ), ...[11]

An eschatological messiah, the Mahdi was sent from God to redeem the world. Like many such millenarian figures (Jewish, Christian, and Zoroastrian), his appearance was both a warning and a promise: the old order was about to collapse and a new world of justice and prosperity was soon to appear. Although the exact day of his coming was unknown, it would not remain a secret. Its signs were either chronological or astronomical: the turn of a century or a millennium or the appearance of a comet, an eclipse, or a conjunction.

9 Alice C. Hunsberger, *Nasir Khusraw, the Ruby of Badakshan* (New York: I. B. Tauris, 2000).

10 Rizvi, *Muslim Revivalist Movements*, 69–70.

11 Ibid., 68.

In Islamic history the Mahdi has usually been interpreted in Shiite terms. For both Ismailis and Imamis, he was the hidden Messiah, the savior who at God's call would return to redeem the world. This definition, however, is too limiting, and through the centuries – from the late seventh through the early twentieth – eschatological extremists of many persuasions (both Sunni and Shiite) appeared and reappeared, raising armies and calling for the overthrow of the ruling order.

The other millenarian figure was the Mujaddid (Renewer). According to one tradition Muhammad said: "... at the beginning of each century, God will send a man, a descendant of his family, who will explain [or renew] the matters of religion ..."[12] Unlike the Mahdi, the Mujaddid was primarily a religious figure, called to reform or renew the corrupted tradition. Over the centuries various claimants to the title appeared – rulers, teachers, scholars, and sheikhs. While the turn of both the eighth (800 AH/1397 CE) and the ninth Islamic centuries (900 AH/1494 CE) offered opportunities to many self-styled critics and reformers, the arrival of the millennium (1000 AH/1591–1592 CE) was an event of a different order, calling for a Mujaddid of more than ordinary charisma and authority.

SAFAVID EMPIRE

The folk Islam of fourteenth and fifteenth century Iran was a heterodox mix of beliefs: in saints, miracles, the Mahdi, and the approaching apocalypse. In such an atmosphere there was no shortage of millenarian prophets and movements.

Fazallah Astarabadi (1340–1394), the founder of the Hurufi Sufi Order, was a mystic and a dream interpreter. In an early dream he discovered that he was the *Sahib-i Zaman* (Lord of the Age), an Ismaili epithet for the Mahdi, and a manifestation of the divine. Soon after Fazallah began to travel and proselytize. After he met (but could not convert) Timur, Miran Shah (the conqueror's son) took him into custody and had him executed. "*Huruf* (letters)" referred to Fazallah's belief that God had revealed himself in his word. For the Hurufis, however, the message of the Quran was not easily interpreted. Only by subjecting the text to a secret system could its true meaning be divined.[13]

[12] *Encyclopaedia of Islam,* 2d ed., s.v. "Mujaddid."

[13] *Encyclopaedia of Islam,* 2d ed., s.v. "Hurufi"; B. S. Amoretti, "Religion in the Timurid and Safavid Periods," in Peter Jackson, ed., *Cambridge History of Iran: vol.* 6, 610–56.

A second millenarian prophet of fifteenth century Iran was Muhammad b. Abd Allah Nurbaksh (1392–1464), the founder of the Nurbakshiyya Sufi Order. Born in Kuhistan, he claimed to be a descendant of Musa al-Kazim, the seventh Ismaili Imam. Soon after joining the Kubrawi Sufi Order, Nurbaksh declared himself the Mahdi and began to recruit a military force. Before he reached the battlefield, however, he was captured by the Timurid ruler Shah Rukh (1405–1447) and returned to Herat in chains. After his release he continued to preach and recruit but, following the death of Shah Rukh, he retired to a village near Rayy. Nurbaksh's son, Shah Qasim (d. 1511), succeeded him and found favor with Shah Ismail. The following two sheikhs of the order, however, were perceived as both spiritual and military threats by Shah Tahmasp, and when the last of Nurbaksh's descendants was executed in 1537 the order was effectively extinguished in Iran. In Kashmir the Nurbakshiyya continued to flourish until the end of the sixteenth century.[14]

An earlier millenarian prophet of fourteenth- and fifteenth-century Iran was Sheikh Safi al-Din Ardabili (1252–1334). A pious ascetic and, according to his followers, a miracle worker and a world conqueror, he founded the Safaviyya Sufi Order. Although he attracted a large party of followers, Sheikh Safi al-Din did not appear to harbor any military or political ambitions. Not until the time of Sheikh Junaid (1447–1460), Shah Ismail's grandfather, did the identity of the order begin to change – from an organization of pious mystics into a millenarian movement with political ambitions. From this point onward the Safaviyyah sheikhs were worshipped as saviors and incarnations of god. Junaid gave his Turkish tribesmen military training and a new *ghazi* (religious warrior) ethos. That the reconfigured order was a military success can be seen in Juanid's warm reception at the court of Uzun Hasan (1453–78), the powerful ruler of the nearby Aq Quyunlu state.[15]

Sheikh Haidar (1459–1488), Junaid's son and Ismail's father, acted more like a secular prince than a religious leader. Under him the Safaviyya *ghazi* warriors enjoyed great success on the battlefield. As confirmation of the order's increasing political and military importance, Uzun

[14] *Encyclopaedia of Islam*, 2d ed., s.v. "Nurbakshiyya"; Shahzad Bashir, *Messianic Hopes and Mystical Visions: the Nurbakshiya Between Medieval and Modern Islam* (Columbia: University of South Carolina Press, 2003); Shahzad Bashir, "After the Messiah: The Nurbakhshiyya in Late Timurid and Early Safavid Times," in Andrew Newman, ed., *Society and Culture in the Early Modern Middle East, Studies on Iran in the Safavid Period* (Leiden: Brill, 2003), 295–314.

[15] Roemer, "The Safavid Period," in Peter Jackson, ed., *Cambridge History of Iran: 6*, 200–5.

Hasan gave his daughter in marriage to the young sheikh. An external sign of the continued transformation of the order was the twelve-pointed red cap – the Taj-i Haidar or Haidar Cap – which caused the group's enemies to nickname them "Qizilbash" or "Red Head." Although Haidar was primarily a military commander, his followers accorded him the highest reverence. In the folk Islam of the day Junaid was god, and Haidar was the son of god.[16]

It was Shah Ismail (1501–1524), the last sheikh of the Sufi order and the first ruler of the Safavid dynasty, who finally fused the two roles – religious leader, on the one hand, and military and political ruler, on the other. Shah Ismail seemed to have adopted more fully (or at least to have articulated more clearly) the messianic identity of his predecessors. In his poetry he claimed to be the Mahdi or Sahib-i Zaman, a reincarnation of Ali, and a manifestation of God. While Ismail's early victories over the Aq Quyunlu and his establishment of the Safavid state seemed a confirmation of divine favor, his defeat by the Ottomans at Chaldiran in 1514 appeared to shake the confidence of many Qizilbash. Although the remaining Shahs of the sixteenth century promoted Imami Shiism as the state religion, Tahmasp continued to be worshipped as god while Abbas I never completely lost the respect due the divinely sanctioned leader of a mystical order.

By the middle of the sixteenth century the Hurufi and Nurbakhshiyya Sufi orders had lost their earlier dynamism and eschatological fervor while the Safavid rulers had, for the most part, become political rather than religious leaders. However, with the end of the sixteenth century rapidly approaching a new charismatic leader of an older movement appeared on the scene. The Nuqtavi Sufi Order had been founded by Mahmud Pasikhani (d. 1428). Born in the Caspian province of Gilan, he was initially a follower of Fazallah Astarabadi. Expelled from the Hurufis for arrogance, he lived a life of piety and asceticism – making several pilgrimages to the holy cities of Mecca and Medina and never marrying. In 800 AH/1397 CE he declared himself the Mahdi, the Mujaddid (Renewer) of the new century, and the Padshah-i Hazarat (King of the Millennia). He claimed to be a reincarnation on a higher plane of both Ali and Muhammad, and his disciples called him the Sheikh-i Wahid (Unique One) and the Insan-i Kamil (Perfect Man).[17]

[16] Ibid., 206–9.
[17] Babayan, *Mystics, Monarchs, and Messiahs*, xxxiv–liii, 3–4, 10–35, 19–20, 67; *Encyclopaedia of Islam*, 2d ed., s.v. "Hurufi."

Mahmud wrote sixteen books and 1,001 treatises but none has ever been fully published.[18] He rebelled against *taqlid* (simple adherence to Islamic tradition) and preached the need for *tajdid* (religious renewal and reform). His doctrines were a queer mixture of the philosophical and the religious: There would not be a day of judgment, creation was eternal, and the Quran was the work of Muhammad. He believed in a materialist theory of transmigration. The particles of the body returned to the earth, reemerging later in plant, animal, or human form. Clearly influenced by Zoroastrian theories, Mahmud and his followers sketched out a world year of sixty-four thousand years divided into four cycles of sixteen thousand years each. Each of the four was further divided into an eight-thousand-year Arab (lunar) epoch and an eight-thousand-year Persian (solar) epoch. At one-thousand-year intervals the epochs switched and the guidance of humanity reverted from a "perfected Arab messenger" to a "perfected Persian preacher." The emergence of Pasikhani foretold the end of the one-thousand-year Arab epoch of Islam, and the beginning of the one-thousand-year Persian, Nuqtaviyya epoch. Pasikhani was also influenced by Abu Mashar, predicting that a great revolutionary leader would emerge in 990 AH/1582–1583 CE, ushering in a new Iranian millennium. His most devoted disciples (the *wahid*) were also celibate and were considered capable of achieving unity with the Godhead while his married followers (the *amin*) were not as spiritually advanced.

Although Pasikhani and the Nuqtavis do not seem to have been very popular during the fifteenth century, by the reign of Shah Ismail they had begun to reappear. They are first mentioned among the disciples of Shah Tahir, an Ismaili Imam who had gathered a large following in the village of Anjudan near Kashan. Opposing the heterodox belief of the Ismailis and the Nuqtavis, Ismail dispatched a troop of cavalry to Anjudan. Many of Tahir's followers were killed, and he himself fled to India (1521–1522).

Despite this early persecution, the Nuqtavi movement was not extinguished. It recovered its footing under Ismail's successor, Shah Tahmasp, and gained adherents in Qazvin, Shiraz, and Isfahan. Like the Safavids, the Nuqtavis were Sufis, but they were not organized into brotherhoods, with rules, robes, and a hierarchy. Rather, the Nuqtavi leaders of the sixteenth century were *qalandars*, independent wandering ascetics who attracted

[18] Kathryn Babayan has provided a masterly interpretation of the Nuqtavi movement in Safavid Iran. Although I have adopted a slightly different chronology for the Shah Abbas–Nuqtavi episode, the remainder of my account depends heavily on her thorough, sensitive treatment. *Mystics, Monarchs, and Messiahs*, ch. 4. See also *Encyclopaedia of Islam*, 2d ed., s.v. "Nuqtawiyya."

disciples wherever they went. Dervish Khusrau, son of a well-digger and the Nuqtavi leader who reignited the movement in the mid-sixteenth century, had studied Nuqtavi beliefs and practices in Kashan. Moving to Qazvin, the Safavid capital, in the late 1550s or early 1560s, he took up residence in a neighborhood mosque and quickly attracted a large and devoted following. His success alarmed the Imami religious leaders, and Tahmasp summoned him to court. Khusrau gave evasive answers to questions about his religious beliefs but he abandoned the mosque and, for the remainder of Tahmasp's reign, was a model believer – studying jurisprudence with the ulama and regularly attending Friday prayers.[19]

Although we do not know the exact date of Khusrau's interrogation, it likely took place in the mid-1560s, as the sect became increasingly popular. In fact, the unstructured nature of the Nuqtavi movement is reflected in the divergent treatment of its leaders. In 1565 Shah Tahmasp arrested two prominent Nuqtavis (Abdul Qasim Muhammad Kuhpaya Amri Shirazi, a poet and court official, and his brother, Maulana Abdul Turab, a court calligrapher). Abdul Qasim had been the supervisor of the waqfs (dedicated properties) for the shrines of Ali and Husain in Najaf and Karbela, and, in addition to his Nuqtavi beliefs, had hinted at further heterodox tendencies by writing poems praising two Ismaili Imams. The Shah ordered both men blinded and banished from court. In 1573–1574 Tahmasp attacked the Nuqtavi-Ismaili community of Anjudan and put to death the Ismaili Imam, Murad Mirza. And, in 1576, the last year of his reign, the emperor jailed a Nuqtavi faction in Kashan that was accused of participating in an Ismaili revolt in nearby Anjudan.[20]

Between the death of Tahmasp in 1576 and the crowning of Shah Abbas in 1587 Safavid Iran was wracked by political and religious turmoil. The two weak rulers of the period – Ismail II (1576–1577) and Muhammad Khudabanda (1578–1587) – struggled against the military threats of the tribal Qizilbash chieftains while Ismail II attempted to replace Imami Shiism with the more widely followed Sunni confession. In 1580–1581 a millenarian dervish, claiming to be a reincarnation of Ismail II, gathered a force of twenty thousand men and briefly threatened Muhammad Khudabanda's rule.[21] Although the Nuqtavi community in Kashan was persecuted by the authorities in 1586, in Qazvin Dervish Khusrau not only

[19] Ibid.; Munshi, *Tarikh-e Alamara* 2: 646–7; Babayan, *Mystics, Monarchs, and Messiahs*, 91ff.

[20] For a complete discussion, see Babayan, *Mystics, Monarchs, and Messiahs*, 93–8; *Encyclopaedia of Islam*, 2d ed., s.v. "Nuqtawiyya."

[21] Newman, *Safavid Iran*, ch. 3.

survived but expanded his following. Perhaps the political instability combined with the approaching end of the millennium rendered the urban populace especially receptive to the Nuqtavi message. In any event Khusrau had to build a much larger meeting hall near the mosque, and his kitchen was soon feeding more than two hundred people a day.[22]

During the first three years of his reign (1587–1590), the young Shah Abbas seemed captivated by the Nuqtavi leader. Although he is said to have been initiated into the order as an *amin*, Iskandar Munshi, perhaps feeling the need to disguise the emperor's fascination with such a controversial figure, stated that Abbas was simply trying to ferret out Khusrau's true beliefs. However, as the Nuqtavis had already hailed Tahmasp as the Mahdi, similar claims about Abbas may have tempted the young, insecure, and embattled ruler. After the Shah's visits, his high-ranking soldiers and officials began to frequent Khusrau's quarters – giving him inlaid daggers and other valuable gifts.[23] Although Khusrau was circumspect, remembering his interrogation by Tahmasp twenty years earlier, his chief lieutenants, Ustad Yusufi the Quiver Maker and Dervish Kucheck Qalandar, were not – apparently trumpeting the group's claim that Khusrau was a semi-divine reincarnation of the Mahdi. On hearing this Abbas was spurred to action.[24]

By 1590 the young ruler had begun his program of political, economic, and religious reorganization. In this new dispensation the millenarian, extremist claims of the Nuqtavi leaders (Dervish Khusrau, Dervish Kucheck, and Ustad Yusufi) had no place. Although Khusrau had elevated both Tahmasp and Abbas to an exalted status in the Nuqtavi pantheon, in the new Safavid state religious legitimacy depended on the emperor's support of Imami Shiism. In 1591, at the turn of the millennium, a Nuqtavi uprising surfaced in Fars and the poet Abdul Qasim Shirazi, twenty-five years after his blinding in Qazvin, was arrested and, at the demand of the ulama, torn to pieces. In 1592 Mir Sayyid Ahmad Kashi (or Kashani), a prominent Nuqtavi leader, was captured and personally beheaded by Abbas. A *firman* (royal decree or order) from the Mughal Emperor Akbar, found among his papers and offering support to Kashi, Dervish Khusrau, and the beleaguered Nuqtavi community, may have added to Abbas's paranoia.[25] The very next year, just before embarking on a military

[22] Munshi, *Tarikh-e Alamara* 2: 647–8; *Encyclopaedia of Islam*, 2d. ed., s.v. "Nuqtawiyya;" Babayan, *Mystics, Monarchs, and Messiahs*, 100–3.

[23] Babayan, 103.

[24] Ibid., 103–5; Munshi, *Tarikh-e Alamara* 2: 648.

[25] Islam, *A Calendar of Persian Documents*, 1: 101–2.

campaign in Luristan, the emperor was warned by Dervish Khusrau that if he didn't return to the capital by the lunar New Year (1 Muharram 1002, 27 September 1593) a Nuqtavi might seize the throne.[26] Soon after Dervish Kucheck issued a similar warning, and, in response, Abbas ordered a detachment of cavalry back to the capital. The horsemen, led by a certain Inayat Kal-i Isfahani,[27] surrounded Nuqtavi headquarters, capturing Dervish Khusrau, Dervish Kuchek, and Ustad Yusufi. Kuchek committed suicide and Khusrau was tortured by the ulama for three days, his dead body left exposed in the streets for an entire week.[28]

Returning from Luristan, the Shah still had to deal with Ustad Yusufi. A clever and persuasive individual (a magician and a Zoroastrian atheist, according to Natanzi), the quiver maker had briefly captivated the young ruler.[29] However, when Jalal al-Din Munajjim Yazdi, the chief astrologer and historian, warned that an inauspicious heavenly event (either a comet or a conjunction of Mars and Saturn) foretold the death of an Iranian ruler, Abbas (on Yazdi's advice) abdicated the throne for the three critical days – 25–28 July 1594.[30] Ustad Yusufi was put in the Shah's place and, after three days in power, was hanged.[31] Although the year-long campaign which followed was the decisive blow, the emperor continued his anti-Nuqtavi campaign for the next fifteen years. In 1600 during a pilgrimage to Mashhad Abbas found his caravan infiltrated by members of the group, and in 1611 he executed the Nuqtavi astrologer Mulla Aya. The last appearance of the sect was in 1631–1632. A certain Dervish Riza, claiming to be the Mahdi, raised a substantial following and seized control of Qazvin. Shah Safi, Abbas's successor, sent imperial troops to recapture the city. Riza was beheaded and his followers executed.[32]

Although the Nuqtavis represented the largest and most successful millenarian movement of sixteenth century Iran, Abbas's extermination

[26] *Encyclopaedia of Islam*, 2d ed., s.v. "Nuqtawiyya"; Munshi, *Tarikh-e Alamara* 2: 648–9. Natanzi wrote that it was Ustad Yusufi who issued the warning. Babayan, *Mystics, Monarchs, and Messiahs*, 105–6.

[27] Charles Melville, "New Light on the Reign of Shah Abbas: Volume III of the Afzal al-Tawarikh," in Newman, ed., *Society and Culture in the Early Modern Middle East*, 83.

[28] Ibid.

[29] Babayan, *Mystics, Monarchs, and Messiahs*, 67.

[30] Munshi, *Tarikh-e Alamara* 2: 648; *Encyclopaedia of Islam*, 2d ed., s.v. "Nuqtawiyya."

[31] While Babayan dated the abdication to the summer of 1593 (1001 AH) both Iskandar Munshi and Hamid Algar had it a year later (1002 AH, 1594). Babayan also described the astrological event as a rare conjunction of Jupiter and Saturn, whereas Iskandar Munshi's conjunction was of Mars and Saturn. Babayan, *Mystics, Monarchs, and Messiahs*, 3, 350; Melville, supra, agrees with Algar in the *Encyclopaedia of Islam*, 2d ed., s.v. "Nuqtawiyya."

[32] *Encyclopaedia of Islam*, 2d ed., s.v. "Nuqtawiyya."

of the group did not mean the elimination of eschatological speculation. For example, the historian Qazi Ahmad, writing at the turn of the millennium, described Sheikh Safi al-Din as the Mujaddid of the seventh Islamic century (700 AH/1300–1301 CE).[33] And Abbas himself had dedicated the magnificent congregational mosque (the Masjid-i Shah) in his new capital of Isfahan, to the Mahdi.[34] Thomas Herbert visited Iran at the end of Abbas's reign and commented on "... [the] books they read, namely Hippocrates, Galen, Avicenna, and *Albu-Mazar* [Abu Mashar]."[35] And in the 1660s Jean Chardin described a stable in the imperial palace which "... they call the Tavila-i Sahib-i Zaman ... the stable of the king of the age ... and they maintain perpetually ... the horses ... one for him [the Mahdi] and one for Jesus."[36]

MUGHAL EMPIRE

During the sixteenth century, early modern India witnessed the appearance of several millenarian prophets.[37] Feeding on the religious, social, and political uncertainty that accompanied the establishment of the Mughal Empire and the end of the millennium, these men built substantial followings. Two of them founded Sufi orders among the Afghan tribesmen of the north Indian borderlands. In 1569 Mulla Muhammad, an Afghan from Baluchistan, proclaimed himself the Mahdi and founded the Zikri Sufi Order. Claiming a spiritual position higher than the prophet, he said: "There is no God but God and Nur Pak [Mulla Muhammad], the Mahdi, is the prophet of God." The Zikris rejected the outward observances of orthodox Islam and concentrated on *zikr* (repetition of the name of God). In their houses of worship (called Zikrana) men and women worshipped together. Accused by the orthodox of un-Islamic innovations, they were persecuted by the Mughal authorities.[38]

Miyan Beyazid Khan (1525–1585), known as Pir Raushan, founded the Raushaniyya Sufi Order. Declaring himself the Mahdi, he developed a large following among the Afghan tribesmen of northwest India. He

[33] Quinn, *Historical Writing*, 81–3.
[34] Blake, *Half the World*, 144.
[35] H. J. Winter, "Persian Science," in Jackson, ed., *Cambridge History of Iran: 6*, 585.
[36] Chardin, *Travels*, 6: 457.
[37] For an interesting discussion of some of these issues, see Ahmed Azfar Moin, "The Islamic Millennium in Mughal India: An Historiographical Analysis," (M.A. thesis, University of Texas at Austin, 2005).
[38] Nizami, *Akbar and Religion*, 69.

preached pantheism, the transmigration of souls, and the equality of all believers. At first Akbar took a favorable view of Beyazid Khan and his order. He sympathized with their heterodox beliefs and thought they would weaken the Afghan opposition. However, when the Raushanis began to rob travelers and disrupt trade, Akbar ordered the governor of Kabul to arrest Beyazid. After the ulama failed to refute his arguments about himself and his mission, he was released. Miyan Beyazid was succeeded by his son, Sheikh Umar, but Akbar, still suspicious, defeated his warrior followers in the last decades of the sixteenth century. The military threat of the movement, however, was not finally extinguished until the reign of Jahangir. The ordinary tribesmen were sent home and their leaders were offered high ranks in the *mansabdari* system.[39]

A third millenarian prophet was Sayyid Muhammad of Jaunpur (1443–1505). In early life a pious and charismatic disciple of the eminent Chishti master, Sheikh Daniyal, Sayyid Muhammad won a large and devoted following. In 1489 he left Jaunpur for Mecca. After circumambulating the Kaba, he declared himself the Mahdi. In Ahmadabad, after his return, he said he had received a message from God: "You are the promised Mahdi; proclaim the manifestation of your Mahdiship and do not fear the people."[40] His movement, called the Mahdawiyya, grew quickly, antagonizing both the ulama and the authorities, and he was forced to leave Gujarat. He travelled through northwest India and Afghanistan to Khurasan, where he died in 1505.

Although the Mahdawi movement grew in size and militancy after the founder's death – mostly among the Afghan tribesmen of the northwest – Jaunpuri was not proposing new beliefs or practices. Unlike Mulla Muhammad or Miyan Beyazid, he intended only to purify the degraded Islam of his day, returning the community to the pristine purity of the early seventh century.

After Sayyid Muhammad's death, his Gujarati followers reorganized themselves into small self-contained, Sufi-like communities called *dairas* (circles). Governed by a rigid interpretation of Islamic law, these early Mahdawis spent their lives in prayer and meditation. A series of deputies succeeded Sayyid Muhammad but, by the middle of the sixteenth century, persecuted by the religious and political elite, the Mahdawi leaders began

39 Ibid., 61–9.

40 Derryl N. MacLean, "Real Men and False Men at the Court of Akbar," 199–215, in David Gilmartin and Bruce Lawrence, eds., *Beyond Turk and Hindu: Rethinking Religious Identities in Islamicate South Asia* (Gainesville: University Press of Florida, 2000), 200. Nizami, *Akbar and Religion*, 41–57.

to preach opposition to the Gujarati Sultan. The second deputy wrote: "It has now become a general religious duty . . . for all – men, women, slaves, and freemen – to unite and defeat the oppressors so that the faithful might be victorious."[41]

Although Gujarat and Sind remained the principal centers of Mahdawi popularity in the mid-sixteenth century, Biyana in north India, near the Mughal capital of Agra, became the home of two important leaders. Sheikh Abdullah Niyazi (d. 1590), a disciple of Sheikh Salim Chishti (1478–1572), was sent on a pilgrimage to Mecca. On his return he met the leader of the Mahdawi community in Ahmadabad and was converted. In Biyana his zeal transformed the small group of believers into a large and vibrant movement. In 1548–1549 the Afghan ruler of north India, Islam Shah Sur (1545–1554), summoned Niyazi to court. Convinced that the Mahdawi was a heretic and a supporter of tribal insurgents, he had him tortured. Niyazi recovered but left Biyana for the village of Fathpur, where Sheikh Salim Chishti had his headquarters. Some twenty years later Akbar's men erected the Ibadat Khana on the site of his abandoned hut. In 1585 Akbar met Sheikh Abdullah near Sirhind and offered him a pension but in 1591, at the turn of the millennium, he died.[42]

The second Mahdawi leader in Biyana was Sheikh Alai. A disciple of Abdullah Niyazi, he established a residence near his master and attracted a great following. In 1550, after punishing Niyazi, Islam Shah ordered Sheikh Alai to court. Although he mounted a vigorous defense of his claim to be the Mahdi, he was also beaten. He lingered a few days and then died.

Given the eclectic atmosphere at Akbar's court, it is not surprising to find that the Mahdawis played an important role in the Ibadat Khana discussions. Sheikh Mubarak, the father of Abu al-Fazl, knew the leading Mahdawi scholar of Gujarat, Sheikh Abd al-Malik Sajanwandi (d. 1574) and had attended the lectures of Sheikh Alai. It was, however, the conquest of Gujarat in 1572–1573 that brought the movement and its leaders to Akbar's attention. The Mughals captured Sheikh Mustafa Gujarati (1525–1576), the leading Mahdawi of Gujarat, and decimated his *dairah*. Sheikh Mustafa's father, Sheikh Miyan Abdur Rashid, who had been converted by Muhammad Jaunpuri himself, was killed along with seven others. Sheikh Mustafa was tortured and sent to Fathpur Sikri, where he resided

[41] MacLean, "Real Men and False Men," 201. See also Rizvi, *Religious and Intellectual*, 57; Rizvi, *Muslim Revivalist Movements*, chs. 2–3.

[42] Rizvi, *Muslim Revivalist Movements*, 124–5.

for nearly two years (1574–1576). In the Ibadat Khana sessions he defended Jaunpuri's claim to be the Mahdi and Mujaddid of the ninth Hijra century (1 Muharram 900 AH/2 October 1494 CE).

Mahdawi influence peaked in the last quarter of the sixteenth century. After the defeat of the Gujarati sultan and the incorporation of his state into the Mughal empire, their military importance declined. And after the turn of the millennium their missionary zeal waned. None of the Mahdawi predictions – the return of the Mahdi or the Day of Judgment – came to pass.[43]

Mulla Muhammad, Miyan Beyazid, and Sayyid Muhammad were all Indian Muslims who founded large millenarian movements in early modern India. Although they and their followers were partly responsible for the apocalyptic anxiety among the populace at large, only Sheikh Mustafa Gujarati had much impact on the Mughal court. In fact, a careful look at the Fathpur Sikri debates reveals that the most eloquent and influential preacher of eschatological extremism was a Nuqtavi from Safavid Iran. Like Mir Fathullah Shirazi and many other Iranian artists, merchants, soldiers, and administrators, Sharif Amuli came to Mughal India seeking refuge and fortune. Because, however, he was so divisive and controversial (in both Iran and India), an accurate picture of the man and his ideas is hard to develop. In the Safavid sources he is mentioned only once. Iskandar Munshi, writing about Abbas's campaign against Dervish Khusrau, described the shah's beheading in 1592 of Mir Sayyid Ahmad Kashi. Among Kashi's papers was the *firman* from the Mughal emperor Akbar, and this led Munshi to recall "Sharif Amuli, who was one of the leading intellectuals of the Nuqtavi sect, [who] fled to India to escape retribution at the hands of the judges. He was honored by the Mogul emperor, emirs, and nobles, who treated him as their spiritual adviser."[44] Reminded by Akbar's *firman*, Iskandar Munshi supplied the only information he had on the most influential Safavid Nuqtavi in India. From the more abundant Indian evidence, however, it seems clear that Munshi's memory was faulty – Amuli actually met Akbar in 1577, about fifteen years before Kashi's beheading.

The Mughal sources on Sharfi Amuli are richer but also uneven. Although several authors mention him in passing, the most complete account was by Akbar's chief critic, Abd al-Qadir Badauni. Badauni's complaint was not so much that Amuli made apocalyptic claims for both

[43] Ibid., 134.
[44] TAAA 2: 650.

Pasikhani and himself but that he was a Shiite, even worse an Ismaili Shiite. According to Badauni, Sharif Amuli fled Iran for Balkh and the compound of the Sufi master Mir Muhammad Zaid.[45] Although Iskandar Munshi revealed nothing about Amuli's earlier life, he was probably caught in Shah Tahmasp's roundup of Nuqtavi leaders in the mid-1560s – when Dervish Khusrau was questioned and the two brothers (the poet Abdul Qasim and the calligrapher Abdul Turab) were blinded. Forced out of Balkh (because his heretical views made him unwelcome), he arrived in the Deccan, probably in the early 1570s. Although we know almost nothing about Amuli's south Indian sojourn, it is tempting to recall that Afzal Khan, the chief minister of Bijapur, had invited Mir Fathullah Shirazi to the Deccan kingdom in the 1570s. Perhaps Amuli, an Iranian Shiite also, was part of the Safavid influx. In any event, he eventually became unpopular in the Deccan as well (because of "the filthiness of his disposition"). In 1577 (the first solid date available) he turned up at the camp of the emperor Akbar.[46]

Unlike Sheikh Mustafa Gujarati or Mir Fathullah Shirazi, however, Sharif Amuli had a long and successful career in Mughal India, spanning nearly thirty years. During the Fathpur Sikri period (1572–1585), he was an intimate of Akbar. Contributing to the wide-ranging discussions in the Ibadat Khana, he joined Akbar's new imperial order (Tauhid-i Ilahi) and supported the principle of "lasting reconciliation." After the emperor left Fathpur Sikri in 1585, Amuli shifted his focus from court to career. In 1586 he was appointed chief religious official of Kabul, in 1587 he was made revenue administrator and distributor of charities for Kashmir, and in 1591 he was raised to the rank of 1000 *zat* and given three posts in the rich province of Bengal: chief judge and head of both the revenue and religious departments. In 1598 the revenues from the province of Ajmir were assigned to him. In 1605, after Akbar's death, Jahangir promoted him first to the rank of 2500 *zat* and then several years later to his final rank of 3000 *zat*.[47]

Amuli's opinions, as reported by Badauni, leave little doubt as to his Nuqtavi sympathies. In 1577 the loud cries of the Iranian's disciples, proclaiming him the Mujaddid of the Millennium, reached Akbar's ears, and he invited him over. Rather like Dervish Khusrau's first meeting with Shah Abbas (about ten years later), Amuli immediately won over the young emperor – with a similar combination of charisma and doctrine.

45 Badaoni, *Muntakhab*, 2: 253.
46 Ibid., 2: 252–5.
47 Rizvi, *Socio-Intellectual History* 2: 192–6.

Taken into the emperor's retinue, he returned to Fathpur Sikri where he remained for the next eight years.[48]

In the last part of 1580 Amuli and his followers began to publicize the astrological predictions of Abu Mashar.

> In this year [November or December 1580] low and mean fellows, who pretended to be learned, but were in reality fools, collected evidence that his Majesty was the Cahib-i-Zaman [Sahib-i Zaman] ... Sharif [Amuli] brought proofs from the writings of Mahmud of Basakhwan [Pasikhan] that he had said that in the year 990 [1582] a certain person would abolish lies.... And Khwajah Maulana of Shiraz ... came with a pamphlet by some of the Sharifs of Makkah, in which a tradition was quoted to the effect that the earth would exist for 7000 years, and as that time was now over the promised appearance of the Mahdi would immediately take place.... The Shiahs mentioned similar nonsense connected with Ali, and quoted the following Ruba'i, which is said to have been composed by Nacir-i Khusrau ...
>
> In 989 [1581–1582], according to the decree of fate,
> The stars from all sides shall meet together
> In the year of Leo, the month of Leo, the day of Leo
> The Lion of God will stand forth from behind the veil.
>
> And this made the Emperor more inclined to claim the dignity of a prophet, perhaps I should say the dignity of something else.[49]

This passage nicely illustrates the several strains of millenarian expectation in late-sixteenth-century India. Amuli's reference to Akbar as Sahib-i Zaman (Lord of the Age) recalls the specifically Ismaili epithet for the Mahdi. And the reference to Pasikhani and 990 (1582), of course, refers back to the Nuqtavi founder's appropriation of Abu Mashar. Khwajah Maulana, the Iranian Shiite and perhaps a Nuqtavi as well, has slightly rewritten Pasikhani's teaching about alternative eight-thousand-year Arab and Persian epochs and the arrival of a Mahdi at the end of each one. The passage is tied together by the quatrain from Nasir-i Khusrau, referring to Abu Mashar's Grand Conjunction, and the final sentence hints at Akbar's interest in assuming the millenarian mantle of the Mahdi – standing "forth from behind the veil."

Sharif Amuli's predictions – of the conjunction and the coming of the Mahdi – were not easily forgotten. Several months later (February or March 1581) Akbar had a conversation with Father Antonio Monserrate, a Portuguese Jesuit. The emperor wondered about

48 Badaoni, *Muntakhab* 2: 252–5.
49 Ibid., 2: 295.

... the last Judgment, ... [whether] Christ would be the Judge, and when it would occur ... the Priest said "God knows the time. ... Yet signs shall precede the day which will enable men to conclude with confidence that it is at hand. ... The king asked what these signs should be. The Priest replied, "Christ mentioned especially wars and rebellions, the fall of kingdoms and nations, the invasion, devastation and conquest of nation by nation and kingdom by kingdom; and these things we see happening very frequently in our time.[50]

Some eighteen months later, in the latter part of 1582 (990), Sharif Amuli created another stir. Badauni wrote:

Some shameless and ill-starred wretches also asked His Majesty why, since a thousand years from the Hijrah were passed, he did not bring forward, like Shah Ismail the First some convincing proof.... The following quatrain of Nasir-i Khusrau was often quoted at court:

> I see in 992 [1584–85] two conjunctions
> I see the sign of the Mahdi and that of [the] Antichrist;
> Either politics or religion must change,
> I clearly see the hidden secret.[51]

Because the translator omitted the line introducing Nasir-i Khusrau and changed the date from 992 to 990, the importance of this passage has not been fully appreciated. And, while Badauni did not identify the "shameless and ill-starred wretches" there is little doubt – given the insults, the conjunction and Mahdi themes, and the identity of the poet – that they were Amuli and his followers.

Amuli's oft-repeated theme – Akbar as divinely appointed leader (Mahdi, Sahib-i Zaman, or Mujaddid) of the new millennium – must have found a sympathetic hearing. In 1584, at the end of the Fathpur Sikri period, the emperor sent the *firman* mentioned earlier to the Iranian Nuqtavi, Mir Sayyid Safi al-Din Ahmad Kashi. It was more a letter from a *murshid* (Sufi master) to a *murid* (disciple) than an order from a prince to a subject, as if Kashi (like Amuli) were a member of the Tauhid-i Ilahi. Akbar wrote that "... he is spiritually close at hand to give succor," and that Mir Sayyid should "... report his spiritual experiences to, and seek guidance from him (Akbar), who is indeed the solver of problems spiritual and universal." Akbar tells Kashi to assure Dervish Khusrau (at the peak of his popularity and influence in Qazvin) of "... royal esteem and favor,"

50 *Commentary of Father Monserrate*, 129.

51 Badaoni, *Muntakhab* 2:323. For the Persian, see Bad al-Qadir Badauni, *Muntakhab al-Tawarikh*, Maulavi Ahmad Ali and W. N. Lees, eds., 3 vols. (Calcutta: Asiatic Society of Bengal, 1865–9), 2: 331. Abu al-Fazl, *Ain* 1: 208.

and to keep him [Akbar] informed of the activities of the members of the Nuqtavi community in Iran."[52] This is the document, found some eight years later, that caused Iskandar Munshi to label Akbar and Abu al-Fazl Nuqtavis and, to name Sharif Amuli as the catalyst for their conversion. Some months later, after receiving Akbar's letter, Kashi responded, declaring that he prayed night and day for the emperor's prosperity, safety, and happiness and that he was ready to perform any service that might be required.[53]

Because Badauni was a jealous and disappointed office-seeker and rabidly anti-Shiite, it is difficult to evaluate his testimony. A perusal of his *Najat al-Rashid*, however, suggests that his criticism of Amuli was the result more of personal prejudice than of theological orthodoxy. Composed in 1591, at the end of the millennium, *Najat al-Rashid* was a religious tract – 531 pages on the vices of the soul, with an exhaustive list of venial and mortal sins. Highly critical of Akbar and the Shiites, Badauni wrote admiringly of both Muhammad Nurbaksh and Muhammad Jaunpuri, complimenting their piety and accepting their claims to be the Mahdi. Nurbaksh's calling was confirmed, he argued, by an important conjunction of Jupiter and Saturn.[54]

Badauni's contemporaries, however, did not share his intense dislike of Amuli. They admired the Iranian's lifestyle, respected his rank, and found his millenarian statements and sectarian affiliation unexceptional. Abu al-Fazl, for example, portrayed Amuli as a Sufi ascetic who had become a reluctant official.[55] For Nizam al-Din Ahmad he was a *mansbdar* of one thousand with an extensive knowledge of mysticism.[56] And Jahangir described the Nuqtavi as a Sufi and a recluse who had achieved high rank.[57] That Amuli's millenarian, apocalyptic characterization of Akbar had become widely accepted can be seen in the first pages of the *Akbar Nama*. After devoting several chapters to the infant Akbar's "nativities" (birth horoscopes), Abu al-Fazl quoted the Persian poet Khaqani (d. 1185):

[52] Islam, *A Calendar of Persian Documents*, 1: 101–2.

[53] Ibid., 1: 103.

[54] A. Zilli, "Badauni Revisited: An Analytic Study of Najat-ur-Rashid," in I. H. Siddiqi, ed., *Medieval India: Essays in Intellectual Thought and Culture*, vol. 1 (New Delhi: Manohar, 2003), 141–66.

[55] I. H. Siddiqui, "Nuqtavi Thinkers at Akbar's Court: A Study of Their Impact on Akbar's Religious and Political Ideas," *Islamic Culture* 72 (1998), 68; Rizvi, *Religious and Intellectual*, 432.

[56] Siddiqui, "Nuqtavi Thinkers," 69.

[57] Rizvi, *Religious and Intellectual*, 432.

They say that every thousand years
There comes into existence a true man ...
Every now and then, the world is saturated with wretches
Then a shining soul comes down out of the sky[58]

During Shahjahan's reign Amuli was remembered as a Sufi, a scholar, and a supporter of Akbar's "lasting reconciliation" policy.[59] It was not until the reign of the conservative Aurangzeb that his extremist views were criticized. Khwaja Kalan, an eminent sheikh of the Naqshbandi order, characterized Amuli as a heretic, a *mansabdar* who had persuaded Akbar to become a Nuqtavi.[60]

The last millenarian Sufi of early modern India was Sheikh Ahmad Sirhindi (1564–1624). As one of the leaders of the Naqshbandi order in the first quarter of the seventeenth century, he was critical of Akbar's tolerance of Hindu beliefs and customs. As a result, sectarian historians of the early twentieth century have tended to portray Sirhindi as the savior of Indian Islam, the man who turned Jahangir and later Aurangzeb away from Akbar's eclectism and toward a more traditional form of belief.[61] While a newer generation of historians has exploded this stereotype, separating the contemporary, nonideological evidence from the later, hagiographical, Sirhindi's claim to be an eschatological revolutionary – the Mujaddid-i Alf-i Sani or Renewer of the Second Millennium – has not been fully appreciated. In a letter to Mir Muhammad Numan, he wrote:

Be it known that a Mujaddid has appeared at the head of every century but the Mujaddid of a century is different from that of a Millennium. Just as there is a world of difference between one hundred and one thousand, so does the Mujaddid of a century differ from that of a Millennium ...[62]

Although the Naqshbandi order flourished under Sirhindi's leadership, its growth spurred by his disciples' insistence that he was the apocalyptic prophet of the second Islamic millennium, among the religious and political elite there was a good deal of skepticism. Just as the enthusiasm for Akbar as Mahdi waned after the turn of the millennium, so the criticism of Sirhindi grew increasingly strident as the seventeenth

[58] Abu al-Fazl, *Akbar Nama* 1: 142.
[59] Siddiqui, "Nuqtavi Thinkers," 70–1.
[60] Rizvi, *Religious and Intellectual*, 493.
[61] Rizvi, *Muslim Revivalist Movements*, chs. 7–8; Y. Friedman, *Sheikh Ahmad Sirhindi: An Outline of His Thought and a Study of His Image in the Eyes of Posterity* (Montreal: McGill University Press, 1971), 77–80.
[62] Rizvi, *Muslim Revivalist Movements*, 264.

century advanced. Sheikh Abd al-Haqq Dihlawi, a friend and an eminent Naqshbandi leader himself, reminded the sheikh that a true Sufi exhibits "self-mortification, humility, courtesy, and politeness" rather than pride and self-promotion.[63] And Jahangir, after questioning Sirhindi and finding him "proud and self-satisfied with all his ignorance," threw him in prison for a year.[64]

After the sheikh's death in 1624 a hagiographical literature sprang up, describing the signs and miracles that validated his claim to be Mujaddid-i Alf-i Sani. In 1679, however, there was another reaction. Aurangzeb ordered the Sheikh al-Islam to prohibit Sirhindi's disciples from spreading false doctrine.[65] And in 1682, probably because Aurangzeb's ban had not been effective, the Indian ulama appealed to the Sharif of Mecca. After examining Sirhindi's writings, the Sharif declared him an infidel.[66] Although the twentieth-century image of Sirhindi as conservative champion of true religion is certainly false, the popular success of his movement suggests that millenarian fervor was widespread in the late sixteenth and early seventeenth centuries.

OTTOMAN EMPIRE

In the Ottoman empire, as in the Mughal and Safavid empires, the fifteenth and sixteenth centuries were a time of millenarian turmoil. Self-styled prophets claiming divine inspiration, like the Nuqtavi leaders in Iran or the Mahdawi sheikhs in India, sparked uprisings in Anatolia and the Balkans. An outbreak in Edirne threatened the reign of Mehmed I (1412–1421),[67] and the Safaviyya sheikhs sowed seeds of dissension among the Turkish tribes of eastern Anatolia. In the first half of the sixteenth century there were the millenarian revolts of Shah Kulu, Celal, Tonguz Oglan, and Kalendar.[68] Celal, for example, was a Safavid visionary who, taking the name of Shah Ismail, claimed to be the Mahdi.[69] At the Ottoman court of Sultan Suleiman (1520–1566) one outcome of the increasing apocalypticism was the

63 Friedman, *Sirhindi*, 88–9.
64 Jahangir, *Tuzuk*, 2: 92–3.
65 Friedman, *Sirhindi*, 94.
66 Ibid., 97.
67 Cornell Fleischer, "Ancient Wisdom and New Sciences: Prophecies at the Ottoman Court in the Fifteenth and Early Sixteenth Centuries," in Bagci and Farhad, eds., *Falnama*, 232.
68 Cornell Fleischer, "The Lawgiver as the Messiah: The Making of the Imperial Image in the Reign of Suleyman," in Gilles Veinstein, ed., *Soliman le "Magnifiqe" et son temps* (Paris: La Documentation française, 1992), 160–1.
69 Shaw, *History*, 1:86.

appearance of a (relatively) new eschatological identity, Sahib Kiran (Lord of the Conjunction). Although this epithet had a venerable ancestry, rooted in the conjunction astrology of Abu Mashar, it became in the mid- to late sixteenth century a characteristically Ottoman imperial title.

Although Sahib Qiran (Persian) or Sahib Kiran (Ottoman Turkish) referred to the conjunctions of Jupiter and Saturn, the common translation – Lord of the (Auspicious) Conjunction – suggested a further meaning. When the phrase was applied to a ruler, a prophet, or a charismatic leader, the astronomical event implied was not an ordinary, garden-variety conjunction (one that occurred every twenty years or so), rather this conjunction was understood to be rare and unusual – either one that occurred every 240 years (at the transition from one zodiacal triplicity to the other) or every 960 years (at a Grand Conjunction).

In the Islamic world Sahib Kiran had long been associated with the great Central Asian conqueror Timur (r. 1370–1405).[70] As the Mughal emperor Jahangir wrote: "In these memoirs whenever Sahib Qirani is written it refers to Amir Timur Gurgan."[71]

Since Sahib Kiran described an individual born at the time of a rare astronomical event, the astrologer, in order to make such an identification, had to draw up a "nativity," a birth chart or horoscope at the time of birth. There was, however, no reliable evidence of the year (much less the hour, day, or month) of Timur's birth.[72] In the sources two different explanations were given. In his memoir (*Malfuzat-i Timuri*) the conqueror wrote:

> A celebrated astrologer waited on me and delivered a plan of my horoscope, stating that at the time of my birth the planets were in so favorable and auspicious a conjunction as certainly to predict the stability and duration of my good fortune and sovereignty; that I should be superior to all the monarchs of the age; ... that I should be the protector of religion, the destroyer of idols, the father of my people, that my descendants should reign for many generations ...[73]

In addition to Timur's own explanation, a second account had the court astrologers bestowing the epithet because of a rare conjunction early in his

[70] E. A. Polyakova, "Timur as Described by the 15th Century Court Historiographers," *Iranian Studies* 21 (1988): 31–44.

[71] Jahangir, *Tuzuk* 1: 5.

[72] Beatrice Manz, "Tamerlane and the Symbolism of Sovereignty," *Iranian Studies* 21 (1988), 113, fn. 33.

[73] S. H. Hodivala, "The Laqab "Sahibqiran-I Sani," *Journal of the Asiatic Society of Bengal: Numismatic Supplement* 35 (1921), 98.

reign.[74] Finally, apart from the astronomical uncertainty, the meaning of the phrase was also vague. The general interpretation was "blessed by heaven, a favorite of fortune," but Rashid-i Din (1247–1318), the historian of the Mongols, stated that it also denoted a famous and powerful ruler, the king of a region or an epoch.[75]

To write about millenarianism in the early modern Ottoman empire is to depend on the work of Cornell Fleischer. In his thorough and insightful study of the historian and courtier Mustafa Ali and in several pathbreaking articles, Fleischer uncovered a hitherto unknown vein of apocalyptic writing, preaching, and rumor, especially during the reign of Sultan Suleiman.

For the Ottomans Sahib Kiran never entirely lost its connection to Timur. Mustafa Ali wrote that Timur, in appropriating the title, had laid claim to universal sovereignty, especially over the Ottomans.[76] And though the phrase came to be commonly associated with Suleiman, it had been sporadically applied to earlier Ottoman sultans. Mehmed II, for example, had been so described,[77] and in 1517, following his defeat of Shah Ismail and his conquest of Syria and Egypt, Selim I (1512–1520) was hailed as Muceddid and Sahib Kiran.[78]

But it was during the first half of Suleiman's reign that the influence of millenarian prophets, portents, and epithets reached their peak. Fleischer identified three important figures (and their texts). The first was Abd al-Rahman Bistami (ca. 1380–1455). His *Miftah al-Jafr al-Jami* (*The Key to the Comprehensive Prognosticon*) was a foundational work. A compendium of apocalypses from the fourteenth and fifteenth centuries, it was a major inspiration for the construction of Suleiman's millenarian identity. Bistami was an authority on eschatological themes, and he knew the conjunction astrology of Abu Mashar as well as the art of cabalistic interpretation, although his "hurufi" writings differed radically from those of Fazallah Astarabadi. For Bistami, as for Pasikhani, the world had an age of seven thousand years.[79]

The second author, Mevlana Isa, was an obscure judge. His history, *Cami ul-Meknunat* (*The Compendium of Hidden Things*), in three versions between 1529 and 1543, reflected the apocalyptic tenor of the time.

74 Geo. P. Taylor, "On the Symbol "Sahib Qiran," *Journal of the Asiatic Society of Bengal: Numismatic Supplement* 6 (1910): 575.

75 Polyakova, "Timur," 37.

76 Fleischer, *Bureaucrat and Intellectual*, 278.

77 Fleischer, "Ancient Wisdom," 236.

78 Fleischer, "Lawgiver," 162–3.

79 Fleischer, "Lawgiver," 170; Fleischer, "Ancient Wisdom," 232–6.

In his work he examined the claims of two rulers – Suleiman and the Holy Roman Emperor Charles V – to be the Sahib Kiran, the divinely guided Last World Emperor. The five points of Mevlana Isa's argument, overwhelmingly supporting Suleiman, were summarized by Fleischer:

1. The Muceddid is due by the year 960 AH/1552–1553 CE.
2. In that year will culminate the Grand Conjunction of Jupiter and Saturn, the fourth such since the time of the prophet.
3. In this connection, the epithet Sahib Kiran refers to the universal ruler who will inaugurate the domination of a single religion to coincide with the Grand Conjunction. Other signs point to the nearness of the apocalypse and the identity of Suleiman as Sahib Kiran.
4. There have been twenty-nine hidden saints who have held spiritual sovereignty over the world. Suleiman is the thirtieth and last.
5. The concatenation of signs – the Grand Conjunction, the arrival of the hidden saint, and the nearness of the millennium (1000 AH/1591 CE) – all argue that Suleiman is the millennium leader – either the Mahdi (or Sahib Kiran) or his forerunner.[80]

Although it is not clear how widely Mevlana Isa's history circulated, his text reflects a pervasive apocalyptic atmosphere during the first half of Suleiman's reign. Ibrahim Pasha, Grand Vizier from 1523 to 1536, routinely referred to Suleiman as "Sahib Kiran–i Alem-Penah (Lord of the Auspicious Conjunction and Refuge of the World)" and "Sahib-Kiran–i Rub-i Meskun (Lord of the Auspicious Conjunction and of the Inhabited World)."[81] In the Ottoman archives, petitions, poems, and reports employed Sahib Kiran or Hazret-i Sahib Kirani rather than Padshah to designate Suleiman. And, in two contemporary histories – the *Cihadname-i Sultan Suleiman* of 1529–1530 and the *Suleimanname* of 1540 – he was "Sahib Kiran" and "Sahib Kiran ve Mehdi-yi Ahir uz-Zaman (Lord of the Auspicious Conjunction and Mahdi of the End of Time)."[82]

Mevlana Isa, however, was a judge and his understanding of historical chronology was flawed. His basic error was a confusion of eras – the Hijra and the Jalali. Although historical time in the Ottoman empire was reckoned according to the lunar era, the astronomical time of the stars and planets was solar. The Hijra Era year began at 1 Muharram and was 354 days long, whereas the Jalali Era year began at the Vernal Equinox and ran

[80] Fleischer, "Lawgiver," 165.
[81] Ibid., 166.
[82] Ibid., 168.

for 365 days. The twenty-year conjunctions of Jupiter and Saturn began at the Vernal Equinox (the first point of Aries) and cycled through the twelve signs of the zodiac, returning after 960 (solar) years to the starting point. The significant conjunctions occurred every 240 and 960 years.

The underlying assumption of Mevlana Isa's argument (points one to three) was that a Grand Conjunction had occurred at the founding of the Islamic religion in 0 AH/622 CE. As a result, a second, equally momentous astronomical event was scheduled to occur nine hundred and sixty years later. A Grand Conjunction in 960 AH, the fourth (the other three at two-hundred and forty year intervals) since the time of the prophet, would herald the arrival of the Sahib Kiran. Because the last version of Mevlana Isa's history was dated 1543, 960 AH/1552–1553 was only a decade away, and it was obvious, to the historian at least, that the heavens were declaring Suleiman rather than Charles V the universal sovereign of the apocalypse. Although Mevlana Isa's eagerness to ground his argument in Abu Mashar's work suggests that the astrologer's theory was well known and widely accepted, his misunderstanding of conjunction astrology led him astray. The 960 years between Grand Conjunctions were solar not lunar years, and, because of the eleven-day difference between solar and lunar eras, equaled 990 lunar years. Thus, the first Grand Conjunction after the founding of the religion would be in 990 AH/1582–1583, not in 960 AH/1552–1553.

The third eschatological figure at Suleiman's court was Haydar Remmal (Geomancer), a refugee from Safavid Iran. In a remarkable piece of historical detective work, Fleischer reconstructed the career of this Iranian agent of apocalypticism.[83] Born some time in the late fifteenth or early sixteenth century, Haydar's early training was as a *munajjim*, probably in his native Tabriz. Soon after, however, he began to specialize in the arts of prognostication – the cabalistic interpretation of letters and names, dream interpretation, and other random methods of predicting the future. From his success at the Ottoman court, one can guess that Haydar was a skillful and charismatic individual. He had probably been a favorite of Shah Ismail also, since he was chosen to tutor Tahmasp, Ismail's son and successor. In 1522 Haydar and the young Tahmasp (eight years old) traveled from the Safavid capital of Tabriz to Herat. The geomancer remained in the young

[83] Fleischer has distributed the information on him over several articles. "Seer to Sultan: Haydar-I Remmal and Sultan Suleyman," in Jayne L. Warner, ed., *Cultural Horizons: A Festschrift in Honor of Talat S. Halman* (Syracuse: Syracuse University Press, 2001), 290–9; "Shadows and Shadows: Prophecy in Politics in 1530s Istanbul," *International Journal of Turkish Studies* 13 (2007): 51–62; "Lawgiver as Messiah"; and "Ancient Wisdom." In what follows, I have interpolated and in places added to Fleischer's account.

ruler's entourage as an advisor for the first four years of his reign – 1524–1528. His duties were probably those of a court *munajjim* – casting horoscopes, putting together the annual almanac (*takvim*), and giving advice about when and where to initiate important activities. In 1528, however, he was dislodged from favor. The fanatical Qizilbash tribesmen of the Safaviyya order, intent on converting Iran to Imami Shiism, drove the Sunni Haydar from court.

Although the first evidence of Haydar in Istanbul was an almanac dated 1535, he probably arrived several years earlier – around 1530 perhaps – and remained a trusted advisor to Suleiman for the next thirty years – dying some time between 1559 and 1565. The almanac (*takvim*), a common production of the working *munajjim*, was assembled for the year beginning at the Vernal Equinox (21 March 1535). Although it is impossible to know exactly when Haydar reached Istanbul or what he did in the years before 1535, it is tempting to imagine the charming young *munajjim* specializing in prognostication and building a reputation. The early 1530s were a time of apocalyptic speculation in the Ottoman capital. In 1533, for example, the Hungarian ambassador became aware of two widely circulated rumors. A massive marble lion on the shore of the Bosporus had turned around, formerly eyeing Asia it now faced Europe. The lion had reversed itself two times before: in 1453 from Europe to Asia, foretelling the fall of Constantinople; and in 1526 from Asia to Europe, predicting Suleiman's victory over the Habsburgs. This last reversal, however, was ominous, prophesying a disastrous defeat for the Ottoman forces. The second rumor concerned the vision of a prominent Ottoman official. Two warriors, one Ottoman and one European, appeared in the night sky, proclaiming in unison a victory for the Europeans.[84] In such a fearful, unsettled atmosphere, word of Haydar's prowess may have prompted Suleiman to invite him to court.

In any event, in the compositions that survived – the 1535 almanac, a Persian work of 1536, and a 1538 report – Haydar added additional apocalyptic elements to the Sahib Kiran identity found in the court petitions and documents and in Mevlana Isa's history. Educated in Safavid Iran and steeped in the eschatological imagery of Ismail's court, Haydar maintained that in every age there were two divinely designated sovereigns, one temporal (Sahib Kiran) and one spiritual (Sahib Zaman). Muhammad had combined the two identities, and Suleiman, as the Qutb of the present age, had also united the two. He is the Ruler of the Last Age

[84] Robert Finlay, "Prophecy and Politics in Istanbul: Charles V, Sultan Suleyman, and the Hapsburg Embassy of 1533–34," *Journal of Early Modern History* 2 (1998): 18–20.

and will reign until 990 AH/1582–1583 CE or until 1000 AH/1591–1592 CE.[85] Haydar's addition of Sahib Zaman (the Ismaili epithet for the Mahdi) to Suleiman's millenarian identity suggests a Safavid influence, and the two dates recall Abu Mashar. In Haydar's Persian text of 1536 he referred several times to the *Kitab al-Qiranat* (*Book of Conjunctions*) as did Bistami in his *Key*. This was, we have seen, the work in which Abu Mashar gave the fullest account of his conjunction theory, linking weighty events of the past (the Flood and the founding of Islam, for example) to Jupiter–Saturn conjunctions while also providing important dates for the future. As a *munajjim*, a *remmal*, and a student of Abu Mashar, Haydar understood conjunction mathematics and was easily able to translate dates from one era to the other. Unlike Mevlana Isa, he had the right date (both lunar and solar) for the first Grand Conjunction after the founding of Islam: Suleiman, as the eschatological Lord of the Last Age, would rule either until the first Grand Conjunction after Islam – in 990 AH/1582–1583, not in 960 AH/1552–1553 – or until the turn of the millennium in 1000 AH/1591 CE.

Fleischer's analysis of Ottoman apocalypticism focused on the reign of Suleiman. According to him, the sultan's imperial image underwent a major change over the nearly fifty years of his rule. During the first quarter century the identity encapsulated in the epithet Sahib Kiran held sway. Over the second twenty-five years, however, Suleiman was transformed into the Magnificent, the creator of the classical empire. No longer the eschatological Lord of the Last Days, he became the lawgiver, the all-conquering general, the munificent patron of the arts.

Although Suleiman's reign was certainly the high point of the sixteenth century, one must not overlook the millenarian developments at the end of the period. The reign of Murad III (1574–1595) encompassed both chronological milestones (1582 or 1583 and 1591), and although there does not appear to have been any widespread effort to transform him into an eschatological messiah, the millenarian significance of his reign was widely recognized. Murad himself was preoccupied with dream interpretation, numerology, and other occult methods of prognostication.[86] He commissioned an Arabic edition of Bistami's *Key*, and he ordered a rare and beautifully illustrated astrological work – *The Book of Felicity* by Muhammad ibn

[85] Fleischer, "Ancient Wisdom," 243.

[86] For more information on Murad, see *Encyclopaedia of Islam*, 2d ed., s.v. "Murad III"; Christine Woodhead, "Murad III and the Historians: Representations of Ottoman Imperial Authority in the Late 16th Century Historiography," in Karateke, ed., *Legitimizing the Order*, 85–98.

Amir Hasan al-Suudi – translated from Arabic into Ottoman Turkish. This manuscript, a detailed account of the twelve signs of the zodiac, was a loose rendering of Abu Mashar's *Kitab al-Mawalid* or *Treatise on Nativities.*[87] In addition, at Taqi al-Din's urging, Murad was the only Ottoman sultan to build an observatory. Given the sultan's interests and Taqi al-Din's expertise, it was certainly no accident that the greatest imperial festival of the early modern period (the circumcision festival of Murad's son Mehmed) was scheduled for the Grand Conjunction of 1582. At one point during the festivities the sultan presented two illustrated copies of *The Book of Felicity* to his daughters, Ayse and Fatima.[88]

In his writings, the historian Mustafa Ali reflected the millennial hopes and fears of Murad's reign. In 1587 he composed *Feraid ul-Vilade* (*Unique Pearls on the Birth*), demonstrating the astrological significance of the year of Murad's birth. Later, as the millennium drew nearer, he wrote *Mirat al-Avalim (Mirrors of the World)*, a work highlighting the astrological signs of the impending apocalypse. During this period, his poetry reflected the millenarian themes of disease, want, and revolution, and in the last volume of his universal history, Ali referred to Mevlana Isa's *Compendium*, citing the dire predictions in Bistami's *Key* and the apocalyptic significance of 990/1582–1583 and 999/1591–1592.[89] The historian also redefined the meaning of Sahib Kiran. According to him it could be properly attributed to only three world conquerors: Alexander, Genghis Khan, and Timur. For Suleiman the correct appellation was Mueyed Min Allah (Succored by God). This title denoted a sovereign never defeated in battle and included Mehmed I, Selim I, and Suleiman.[90]

Interest in the esoteric did not end with the sixteenth century. Under Mehmed III (1595–1602), Murad's successor, Bistam's *Key* was translated into Ottoman Turkish, and in 1683 Cezmi Efendi (d. 1692), Kadi of Belgrade, wrote:

> At this time the conjunction of the planets of Jupiter and Saturn, which according to the calculators of the calendrists, would take place in 1094 [1682–1683], was reckoned to take place on the last day of Jumada II of the same year in the calendar prepared on the basis of Ulugh Beg's *Zij*.[91]

87 Muhammad ibn Amir Hasan al-Suudi, "The Book of Felicity," fol. 33r, http://moleiro.com.

88 Massumeh Farhad and Serpil Bagci, "The Manuscripts," in Bagci and Farhad, eds., *Falnama*, 72.

89 Fleischer, *Bureaucrat and Intellectual*, 151–2; "Ancient Wisdom," 243.

90 Fleischer, *Bureaucrat and Intellectual*, 279–80.

91 Ihsanoglu, "The Introduction of Western Science to the Ottoman World: A Case Study of Modern Astronomy," in Ihsanoglu, *Science, Technology*, 8–9.

Because of the congruence of the two millenarian events – the Grand Conjunction of Jupiter and Saturn and the end of the first millennium – an apocalyptic atmosphere pervaded the three early modern Islamic empires. Although each empire responded to these events in its own way, Safavid Iran provided a common repository of theories and prophets. Abu Mashar was an Iranian, and the Ismailis who popularized and passed on his theories – the author of the *Rasail* and Nasir-i Khusrau – were both from Iran. Haydar Remmal fled the Safavid state during the early years of Tahmasp's reign, and Mir Fathullah Shirazi left his homeland during the shah's last decade. The largely underappreciated figure linking these two experts in Iranian millenarianism was Mir Ghiyath al-Din Mansur Dashtaki.

Both men were astrologers and students (informal or formal) of Dashtaki – Haydar Remmal during his eight years (1520–1528) at the courts of Ismail and Tahmasp and Mir Fathullah during his five to seven years at the Mansurriyah (ca. 1535–1542). When Haydar Remmal was in Tabriz – as Tahmasp's tutor (1520–1524) and as his advisor (1524–1528) – Mir Ghiyath al-Din was the chief intellectual presence at court: the principal religious official of the empire and its preeminent philosopher as well. Given Dashtaki's reputation, Remmal's apocalypticism, and their time together (nearly a decade as fellow courtiers), it is tempting to identify the Mansurriyah scholar and teacher as the source for many of the esoteric themes and ideas in Haydar's Istanbul compositions. Mir Fathullah's encounter with Dashtaki, by contrast, was more formal. After Mir Ghiyath al-Din lost his post as Sadr in 1531, he returned to Shiraz and resumed headship of the Madrasa. And it was there during the last decade of his life that Dashtaki ignited Mir Fathullah's passion for the rational sciences. Unlike Haydar, however, Mir Fathullah remained in Iran and, over the next thirty years of writing and teaching, transformed himself into the leading authority in the rational sciences – Dashtaki's successor. In India, Mir Fathullah's career was shorter than Remmal's but more public. His mastery of the several sciences gained him widespread renown, and he was chosen by Akbar to reorganize the Mughal madrasa curriculum. Amid all of his other activity, however, he remained an esoteric visionary – a Shiite astrologer heavily influenced by Ismail's self-claims, the Zoroastrian ideas of the Azaris, and the conjunction astrology of Abu Mashar.

In addition to Mir Ghiyath al-Din, another feature common to the millenarian culture of the three empires was the use of the epithet Sahib Kiran. Based on the conjunction astrology of Abu Mashar, Lord of the (Auspicious) Conjunction was a title for Safavid and Mughal rulers as well as for Ottoman. In the Safavid histories, documents, and coins the epithet was relatively rare. It had been used in histories of Ismail,[92] and his astrologer, Muhyi al-Din ibn-i Badr al-Din Anari, had written several conjunction studies (*jami al-qiranat*) in which the 1582 event was highlighted.[93] The epithet was also found on the coins of Shah Abbas II (1642–1666) and his successor Shah Suleiman (1666–1694).[94] And, after the fall of the Safavid dynasty the epithet was used for Nadir Shah Afshar (1736–1747).[95]

However, it was the Indian rulers who made the more extensive use of the phrase. Because the Mughals were directly descended from Timur, Lord of the (Auspicious) Conjunction was widely used, employed by nine emperors of the dynasty – Jahangir through Akbar II.[96] For the Mughals, though, a slight variant became increasingly popular. Maktub Khan's chronogram, commemorating the accession of Jahangir, read:

> King of kings Jahangir, a second Timur, Sat in justice on the victorious throne. Success, fortune, victory, pomp and triumph are wrapped around him to serve with joy. This is the day of his accession, when fortune puts its head at the feet of the Sahib-i Qiran-i Sani (Second Lord of the [Auspicious] Conjunction).[97]

It was to Shahjahan, however, that the epithet Second Lord of the Auspicious Conjunction was routinely applied. The contemporary historian wrote:

> . . . it was settled that the auspicious surname [of Shahjahan] was to be Sahib Qiran the Second, because the infidel-destroying sword of that Solomon-like Padishah had driven away the demons of strife and disorder from the face of the earth and there was not only a likeness and a resemblance between his praiseworthy qualities and ways, and those of His Majesty the Great Sahib Quran, but the [numerical

92 Flemming, "Sahib-Kiran und Mahdi: Turkische Endzeiterwartungen im ersten Jahrzehnt der Regierung Suleymans," in Gyorgy Kkara, ed., *Between the Danube and the Caucasus* (Budapest, 1987), 62.

93 Babayan, "The Cosmological Order," fn. 14.

94 Sanjay Garg, "Safavid-Mughal Relations: The Numismatic Evidence" (paper presented at the international conference "Iran and Historiography in Comparative Perspective," University of St. Andrews, September, 2009).

95 Ibid.

96 Taylor, "On the Symbol "Sahib Qiran," 576.

97 "Mutribi" al-Asamm Samarqandi, *Conversations with Emperor Jahangir*, trans. Richard C. Foltz (Costa Mesa, CA: Mazda Publishers, 1998), 30. The letters in the words Sahib Qiran-I Sani yield 1013. To this is added the first letter of Iqbal (fortune's head), which has a numerical value of one, for the Hijra 1014/1605, the date of Jahangir's accession. Lisa Balabanlilar, "Lords of the Auspicious Conjunction: Turco-Mongol Imperial Identity on the Subcontinent," *Journal of World History* 18 (2007): 28.

value] of the word Sahib Qiran . . . was found to be exactly equal to the numerical value of the letters of Shah Jahan.[98]

Because there were at least twelve different ways of calculating the numerical value of Persian words and phrases, it is no surprise to find the two historians (Jahangir's and Shahjahan's) arriving at different totals. These explanations, it is interesting to note, recall the cabalistic interpretations of the Hurufis, the Nuqtavis, Abd al-Rahman Bistami, and Haydar Remmal.

In each empire, a conservative backlash followed the millenarian outbursts of the late sixteenth century. In Iran Shah Abbas's attacks against the Nuqtavis – putting to death Ahmad Kashi, Dervish Khusrau, Dervish Kuchek, Ustar Yusufi, and many others – was the early phase in a wide-ranging campaign to establish Imami Shiism as the official confession of the Safavid state. In India, the religious and cultural eclecticism of Akbar's court – seen in the influence of Sheikh Mustafa Gujarati, Sharif Amuli, and later Sheikh Ahmad Sirhindi – was followed in the mid-seventeenth century by a general repudiation of their doctrines and self-claims. And in the Ottoman empire the messianic claims for Sultan Suleiman (by Mevlana Isa, Haydar Remmal, and others) in the mid-sixteenth century and the eschatological fervor under Murad III at the end of the century gave way to the traditionalist reaction of the Kadizadelis in the middle of the seventeenth century.[99]

[98] Hodivala, "The Laqab "Sahibqiran-I Sani," 99.

[99] Madeline C. Zilfi, "The Kadizadelis: Discordant Revivalism in Seventeenth-Century Istanbul," *Journal of Near Eastern Studies* (1986): 251–69.

6

Conclusion

The Islamic temporal system – in both its early form and its later early modern version – was a radical redefinition of the concept the new community had inherited. To gain a better appreciation of its revolutionary character it is helpful to view it in context, against the backdrop of the Judeo-Christian system.

The Islamic day, like the Jewish, began at sundown. However, the early institution of the five daily prayers, especially the noon prayer on Friday, created an immediate interest in more precisely dividing the day. Although prayer was an important part of daily life for Jews and Christians, it was not so exactly scheduled. For the ordinary Jew, prayer was only a general obligation – "evening and morning and at noon."[1] In the second century CE the *Didache* (an early Christian treatise) admonished Christians to privately recite the Lord's Prayer in the morning, afternoon, and evening. By the early third century the clergy were reminded to assemble the people each morning and evening for prayer services. When Christianity gained legal status under Constantine (325), however, the Sunday service (time unspecified) became the central event of the week, and the daily prayers occurred in the morning (Lauds) and the evening (Vespers). After monastic life became popular in the fifth century, St. Benedict introduced the Benedictine Rule (540). Modeled after the rigid daily schedule of the Roman army, the monks' day was organized around eight prayers: Matins (early morning, still dark), Lauds (very early morning), Prime (first hour), Terce (third Hour), Sext (sixth hour, noon), None (ninth hour), Vespers (when the evening star appeared), and Compline (just

[1] Psalm 55:17 *Revised Standard Version*.

before retiring).[2] The important thing to note, however, about this reordering is that it was restricted to monks in monasteries. In the premodern Islamic world, by contrast, the new daily regime governed the lives of the entire community – not just those of the religious specialists.

The Islamic week, like the day, also followed the Judeo-Christian model. Here, however, the revisions were relatively minor – new names for the seven days and a new peak day.

It was, however, in the redefinition of the month and the year that the new temporal system proved most radical. On this matter, God had spoken directly to the prophet, commanding a new order with a clarity unknown in the other two monotheistic religions.

While Allah ordained a month of 29½ days and a lunar year of 354 days (with no intercalation), the Jewish calendrical tradition was more unsettled, having passed through several phases. Unlike the Islamic, the Jewish calendar was lunisolar – that is, both the month and the year were lunar except that every three years or so an extra month was added to keep the important festivals in rough synchronization with the seasons. The names of the months changed over time. In the Torah (Exodus 12:2), only four months were named, and the first (Aviv, later Nisan) began in the spring, around the time of the Vernal Equinox. During the Babylon captivity (ca. 586–538 BCE), however, the Jews adopted the Babylonian names for the twelve months. Although the Torah did not explicitly mention an extra month, the seasonal requirements of the annual festivals implied a lunisolar calendar, and an extra month was added at the end of the year, before the first day of Nisan.[3]

The Christian month and year, by contrast, were solar. Although its debt to Judaism was unmistakable, the Christian temporal system also borrowed a great deal from the Roman. The early Roman calendar was, like the Jewish, lunisolar but with a four rather than a three-year cycle. By the time of Julius Caesar (100–44 BCE), however, the periodic adjustments had either been neglected or manipulated so that the beginning of the year (1 January) had regressed to the early fall. In 46–45 BCE, therefore, the emperor added ninety days to the calendar, returning New Year's Day and the beginning of spring to their proper places in the seasonal round. The calendar was solar – 365 days with a leap year every four years – and the months were each given a definite length, except for February.

After Christianity became the official religion of the Roman Empire, Constantine made three changes in the Julian calendar: he established a

[2] *Demontreville Jesuit Retreat House Information*, 67–8.
[3] Jones, "Calendar," 572–5.

seven-day week, with Sunday the holy day; he officially recognized the Christian holidays of fixed dates such as Christmas; and he grafted on the movable Easter celebrations tied to the shifting date of Passover.[4]

In 1580, Pope Gregory XIII (1572–1586) established a committee to undertake the restructuring of the Julian Calendar. Because the new calendar contained an eleven-minute error, it gained one day every 134 years. By the fourteenth century, the church had become aware of the problem – primarily because of the inappropriate date provided for Easter. Gregory's committee recommended the revision worked out by the astronomer/astrologer Aloysius Lilius (1510–1576), and in 1582 ten days were dropped from the month of October and the calendar was once again brought into synchronization with the tropical year. The interesting question, however, is the date of the reform. Lily's work was completed in the 1570s and Gregory's committee was established in 1580. Given the contemporary interest in astrology and the number of astrologers at the Papal court, it is hard to imagine that introducing this momentous reform in the year of the Grand Conjunction was an accident.[5]

The Islamic redefinition created an intense interest in both the practical (timekeeping) and the theoretical (astronomical/astrological) aspects of time. The need to more accurately proclaim the times of prayer led Islamic *munajjims* to repeatedly redesign their gnomons and sundials, adding more observations and more towns and cities to their shadow-length tables. A little later, in a further commitment to greater accuracy, they also began to introduce water clocks. The superiority of Islamic technology at this point is nicely captured in the story of Harun al-Rashid's gift to Charlemagne.

As for the theoretical aspect of the temporal revolution, it is enough to note that the observatory as an institution was an Islamic creation. Each of the three great Islamic observatories – Malik Shah's in Isfahan (1072–1092), Hulaghu's at Maragha (1256–1265), and Ulugh Beg's at Samarqand (1417–1420) – was in its day the most advanced in the Eurasian world.

The first European observatory is generally considered to be that set up at Nurnberg by Regiomontanus (ca. 1436–1476). Very little is known about its observational activity, however, and the Kassel Observatory, constructed by Wilhem IV (1532–1592), Landgrave of Hesse, was the

[4] Duncan, *Calendar*, 53.

[5] Eviatar Zeruvabel, *Hidden Rhythms: Schedules and Calendars in Social Life* (Chicago: University of Chicago Press, 1982), 97–9; August Ziggelaar, "The Papal Bull of 1582: Promulgating a Reform of the Calendar," in George V. Coyne, ed., *Gregorian Reform of the Calendar* (Vatican City, 1983), 201–39.

site of the first documented program of astronomical observations in Europe. Although the scientific work of the German observatory probably compared favorably with that at Samarqand a hundred years earlier, it was not until Tycho Brahe (1546–1601) erected the Uraniborg Observatory in 1576 that we encounter a European institution whose work clearly surpassed that of its Islamic contemporaries.[6]

Just as the Islamic observatories surpassed the European in size, funding, and instrumentation so too was their scientific output superior: their astronomical treatises were more precise, comprehensive, and mathematically sophisticated. The most popular astronomical treatise in Europe from the thirteenth century until the early seventeenth was the *Alfonsine Tables* of Alfonso X (1252–1284), ruler of Castille. Finished in 1272, at about the same time as the *Zij-i Ilkhani* of Nasir al-Din Tusi, they were a revision of the Toledan Tables of al-Zargali (1029–1087). However, because the new observations, on which the updating was based, were the work of only two observers using small portable instruments, the *Alfonsine Tables* were nowhere near as accurate or as comprehensive as those produced by Tusi at Maragha. Composed in Spanish, the *Alfonsine Tables* were quickly translated into Latin and became the standard reference for the next three hundred years. As a result, by the early fifteenth century the scientific gap between the Islamic and European worlds had widened considerably. The Europeans were stuck with a thirteenth-century astronomical treatise that was doubly flawed – obviously inferior to the roughly contemporaneous *zij* of Nasir al-Din Tusi, on the one hand, while being clearly surpassed by the newly completed treatise of Ulugh Beg, on the other. It was not until 1627, when Johannes Kepler published the *Rudolphine Tables*, using the heliocentric model of Copernicus and data from the observations of Tycho Brahe, that the *Alfonsine Tables* were everywhere superseded.[7]

For the early Muslims, the ceremonial round was organized according to the lunar Hijra calendar. The four basic rituals all regressed against the seasons. By the early modern period, however, the ceremonial simplicity of the first centuries had been lost. In both the Mughal and Safavid empires solar, seasonal festivals were added to the ceremonial cycle, whereas in the Ottoman empire the addition and subtraction of rituals all took place within a lunar context.

[6] Sayili, *Observatory*, 369–400.

[7] Ibid.

For Jews and Christians, by contrast, the seasonal aspects of religious ritual had always been complex. The Jewish calendar was lunisolar, and thus the principal ceremonies of the liturgical year occurred and reoccurred at approximately the same point in the seasonal round. For Jews, autumn was the high point, with three great festivals oscillating around the Autumnal Equinox (21 September). Rosh Hashanah (New Year's Day) on 1 Tishri (September–October) was the first of the High Holy Days. As the Day of Judgment, it was followed ten days later by Yom Kippur (Day of Atonement), the holiest day of the Jewish year. Five days after was Sukkoth (15 Tishri), the Feast of Booths. While commemorating the forty-year sojourn in the wilderness, Sukkoth was also a harvest festival. The other major ritual of the Jewish year, Passover, also had a strong seasonal theme. Celebrating the escape of the Hebrew people from their captivity in Egypt, it occurred on 15 Nisan (March–April). In the ritual calendar this was the first full moon after the Vernal Equinox (21 March). With Passover, as with Sukkoth, the seasonal and religious messages reinforced one another. The spring festival celebrated freedom and new life for the nation during a season of rebirth and renewal.

The Christian ceremonial cycle rested on even more complex temporal foundations. Whereas Constantine's calendar was solar, Easter, the more important of the two major Christian festivals, was lunar. Jesus of Nazareth and his early followers were Jews, and his execution occurred in Jerusalem during the Passover celebration. According to the New Testament, the Last Supper (the final communal meal for Jesus and his disciples) was a Passover feast (on Maundy Thursday). Jesus was crucified on Friday and, according to the early accounts, rose from the grave on Sunday. The decision of the early church to respect sacred chronology meant that Easter had to be celebrated after Passover. Thus, the defining ritual of the Christian year migrated around the Vernal Equinox (21 March), occurring on the first Sunday after the first day of Passover. And, of course, the Easter message of rebirth and new life was reiterated in the regeneration of the earth that accompanied the arrival of spring. As a result, although the Julian calendar of the early church was solar, the minor feasts and festivals of the first half of the Christian year occurred on a lunar schedule.

The other major festival of the Christian calendar was solar. Although the actual birthday of Jesus of Nazareth was unknown, the early church chose to celebrate it on 25 December – the accepted date of the Winter Solstice (astronomically, 21 December). The seasonal symbolism was important. On this day, the shortest of the year, the sun ended its southern retreat. The heavenly source of light was reborn, and the hours of daylight steadily increased. So, too, for early Christians the birth of the savior heralded new life and new hope. The last half of the Christian year, as a result, was solar, the pre-Christmas feasts dependent on the position of the earth in its annual voyage around the sun.

Allah's prohibition against intercalation had a profound effect on Islamic chronology. As with the calendar and the ceremonial cycle, the era created by the prophet's early followers was radically different from those of contemporary Jews or Christians.

The Jewish era was lunisolar. In a nineteen year cycle, seven years (the third, sixth, eighth, eleventh, fourteenth, seventeenth, and nineteenth) were elongated by the addition of a thirteenth month, creating years of 383–385 days instead of 354. The epoch of the Jewish Era, 1 Tishri 1 AM (Anno Mundi, Year of the World), was 7 October 3761 BCE, commemorating the creation of the world. In the Torah, however, the first day of the year (the ecclesiastical new year) was celebrated in the early spring – 1 Nisan (March–April). After the Babylonian captivity, the beginning of the civil year was moved to the early fall, 1 Tishri or Rosh Hashanah (September–October). This was also the point at which the year number changed. With the conquest of Israel by Ptolemy in 63 BCE, the Jews under Roman occupation began to follow the imperial civil calendar as well. Taxes were paid and government matters were settled according to the revised solar calendar of Julius Caesar.[8]

Christian chronology at the time of Muhammad was even more complicated. For the first two centuries after the reforms of Constantine the era used by the Christians was the Diocletian. Named after the Roman emperor of the day, it was solar and its epoch was 1 January 284. In 525, however, Dionysius Exiguous (c. 470–545), a Romanian monk who lived in Rome, published a table calculating the dates of Easter for the next ninety-five years. Because he did not want to perpetuate the memory of a tyrant who had persecuted the Church, Dionysius Exiguous established a new era: Years of Our Lord Jesus Christ, Anni Domini Nostri Jesu Christi (Anno Domini or AD). Although the Venerable Bede in *The Ecclesiastical History of the English People* (731) used Dionysius's era, the Roman Church does not seem to have adopted it until about the tenth century. Although Dionysius did not specify the epoch of his era, most scholars have taken 1 (there is, of course, no year 0) as the year of Christ's birth.[9] To

[8] Nehr, "The View of Time in Jewish Culture."

[9] *Catholic Encyclopedia*, s.v. "Dionysius Exiguous."

further complicate matters, there was no agreed-on date for the beginning of the year. Under the Romans, 1 January had been established as New Year's Day because it was the time at which the consuls took office. After Constantine's reforms, however, other dates were adopted: the beginning of Advent, Christmas, the Annunciation (or Incarnation), and Easter. In fact, toward the end of the Middle Ages in Europe alone there were at least six different dates for the beginning of the year.[10]

Gregory XIII's reform of the calendar gave birth to a revived interest in chronology. His Proclamation of the Birth of Christ, instituted in 1584 and proclaimed annually thereafter, read:

> The twenty-fifth day of December.
> In the five thousand one hundred and ninety-ninth year of the creation of the world
> From the time when God in the beginning created the heavens and the earth;
> The two thousand nine hundred and fifty-seventh year after the flood;
> The two thousand and fifteenth year from the birth of Abraham;
> The one thousand five hundred and tenth year from Moses and the going forth of the people of Israel from Egypt;
> The one thousand and thirty-second year from David's being anointed king;
> In the sixty-fifth week according to the prophecy of Daniel;
> In the one hundred and ninety-fourth Olympiad;
> The seven hundred and fifty-second year from the foundation of the city of Rome;
> The forty-second year of the reign of Octavian Augustus;
> The whole world being at peace,
> In the sixth age of the world,
> Jesus Christ the eternal God and Son of the eternal Father,
> Desiring to sanctify the world by his most merciful coming;
> Being conceived by the Holy Spirit,
> And nine months having passed since his conception,
> Was born in Bethlehem of Judea of the Virgin Mary,
> Being made flesh.
> The Nativity of our Lord Jesus Christ according to the flesh.[11]

In early modern Europe, as in the early modern Islamic world, millenarian prophets and sects played an important role in religious, political, and social life. Although the apocalyptic spirit of the age has been carefully and imaginatively sketched out in a series of articles by Sanjay Subrahmanyam[12] and Cornell Fleischer, a few additional comments might be helpful.

[10] Zerubavel, *Hidden Rhythms*, 98–9.

[11] *Encyclopaedia Britannica*, 11th ed., s.v. "Martyology."

[12] See Sanjay Subrahmanyam, "Connected Histories: Notes towards a Reconfiguration of Early Modern Eurasia, *Modern Asian Studies* 31 (1997): 735–62; Sanjay

From the twelfth to the mid-seventeenth centuries, astrologers stood at the center of European intellectual and political life. Adelard of Bath (ca. 1080–1152) read the Latin translations of the Arabic philosophers and scientists and, like St. Thomas, tried to reconcile Plato and Aristotle with the Holy Scriptures. In the civil war between King Stephen (r. 1135–1154) and his rivals he used his astrological skills to cast nativities and interrogatories for the credulous sovereign and his generals. In the next century Michael Scott, one of the foremost translators of Arabic works into Latin, performed similar services for the Holy Roman Emperor Frederick II (1215–1250). Later, eminent philosophers and scientists such as the Italian Jerome Cardan (1501–76), the Englishman John Dee (1527–1608/9), and the German Johannes Kepler (1571–1630) advised rulers, generals, and popes on the auspicious times for important undertakings. In England the last king to receive regular astrological advice was Charles II (1660–1685), who consulted Elias Ashmole (1617–1692), one of the founders of the Royal Society, on the proper times to address Parliament.[13]

For these astrologers, the work of Abu Mashar (Latinized Albumasar) was central. His *De Magnis Coniunctionibus* (*On the Great Conjunction*), a translation of *Kitab al-Qiranat* (*Book of Conjunctions*), was the inspiration for a variety of historical schema based on the Jupiter–Saturn cycles.[14] The Italian Pietro d'Abano (c. 1250–1316) divided world history into a Great Week of Seven Eras. Each era, ruled by one of the seven planets, was further subdivided according to Albumasar's theories.[15] In the mathematical section of his *Opus Major* (*Major Work*) Roger Bacon (c. 1214–94) wrote

> ... Albumasar, then, in his book on Conjunctions ... determines that there are three conjunctions of Saturn and Jupiter, namely, a great one, a greater, and the greatest. A great one happens at their conjunction every twenty years ... greater conjunction ... happens every two hundred and forty years and ... has reference to a sect and its change in certain regions. When the conjunction has changed from this triplicity to another one, as from the end of Cancer to the beginning of Aries, it is said to be the greatest one ... and happens every nine hundred and sixty years, and has reference to changes in empires and kingdoms.[16]

Subrahmanyam, "Turning the Stones Over: Sixteenth-Century Millenarianism from the Tagus to the Ganges," *The Indian Economic and Social History Review* 40 (2003): 129–61.

13 Nicholas Campion, *The Great Year: Astrology, Millenarianism, and History in the Western Tradition* (London: Arkana, 1994), 353–4.

14 *Abu Ma'sar on Historical Astrology*, 1: xi–xiv.

15 Campion, *Great Year*, 354–5.

16 Aston, "Fiery Trigon," 163–4.

Although Cardinal Pierre D'Ailly (ca. 1350–1420) criticized Bacon's reliance on astrology, he himself cast the horoscopes of Jesus and Mary. In his chronology, the Jupiter–Saturn conjunctions coincided with the Flood, the giving of the Ten Commandments, the establishment of the kingdom of Israel, and the foundation of the Franciscan and Dominican Orders. They also foretold the coming of the Antichrist. He is, however, best known for his prediction that 1789 would be a year of major political upheaval.[17] Franciscus Florentinus (b. ca. 1420), Dean of the College of Theologians in Florence, believed that the history of religion was regulated by the Jupiter–Saturn conjunctions, and Philip Melanchthon (1497–1560), a close associate of Luther, was a practicing judicial astrologer.[18] Nostradamus (1503–1566), the French astrologer and seer, published his first almanac in 1550. An astonishing success, it was followed by eleven others and in 1555 by *Les Prophecies*, his best-known work. He cast horoscopes and nativities at the court of Charles IX (1550–1574) and predicted, on the basis of planetary movements, several dates for the apocalypse.[19]

In 1583 the French Calvinist Joseph Scalinger (1540–1609) published *De Emendatione Temprum* (*Reform of Chronology*), a work credited with establishing the modern science of chronology. During the sixteenth century there was an extensive literature on the *computus* (the calculation of the date of Easter). Scaliger had collected a great deal of information on calendars and eras – European, Islamic, and Asian – and he produced a timeline that combined three cycles: a twenty-eight-year solar, a nineteen-year lunar, and a fifteen-year indiction. His conclusion: the present era had begun in 4713 BCE and would last 7,980 years.[20]

In England, the impending Great Conjunction of 1583 aroused a great deal of anxiety. According to Elizabethan astrologers, this conjunction had occurred only six times since the creation of the world, and but once since the birth of Christ. They linked Albumasar's prediction with an old prophecy that predicted the end of the world in 1588. Thus, the decade from the late 1570s onward was filled with rumors, fears, and dire prophecies, and the gathering of the Spanish fleet in 1588 only exacerbated the hysteria.[21]

Although the Islamic concept of time that emerged during the early Umayyad period was a radical departure, the calendrical, ceremonial, and

[17] Campion, *Great Year*, 354–6.
[18] Ibid., 357.
[19] Subrahmanyam, "Turning the Stones Over," 136.
[20] Anthony Grafton, *Joseph Scalinger: A Study in the History of Classical Scholarship*, 2 vols. (Oxford: Clarendon Press, 1983–94), 2: 7, 102, 349–50. See also Campion, *Great Year*, 361.
[21] Margaret Aston, "Fiery Trigon."

chronological systems of the early modern Islamic empires were in many ways innovations just as revolutionary.

The Safavid empire was the smallest of the three states and the least diverse. Building on its pre-Islamic imperial past, the Safavids employed the 365-day Zoroastrian calendar – New Year's Day was the Vernal Equinox. The Safavids, however, had a history of astronomical achievement. The three greatest observatories of the Islamic world – in Isfahan, Maragah, and Samarqand – were located in greater Iran, and the three principal solar eras – the Yazdegird, Jalali, and Turkish Twelve-Year Animal – were all Iranian. The ceremonial cycle under the Safavids was the result of a negotiation between the demands of the pre-Islamic past and the legitimation needs of the dynastic present. The decision of the Safavid rulers to redefine themselves – from masters of the Safaviyya Sufi Order to champions of Imami Shiism – meant that Ashura slowly began to supplant Nau Ruz as the emblematic ritual of the dynasty.

Because of the size and decentralized structure of their state and the availability of solar eras, the Safavids did not find it particularly difficult to organize and administer an agrarian tax system. What was difficult, however, was deciding how to date histories, proclamations, and records. Safavid historians and administrators adopted different strategies in their efforts to integrate the lunar Hijra Era with the Jalali and Turkish Animal Eras, trying to ensure that the significant dates – of Nau Ruz, Ashura, accessions, battles, marriages, and births – would be accurate across the several systems. Although the popularity of solar chronologies in Safavid Iran meant that the turn of the Islamic millennium passed relatively unnoticed, the fifteenth and sixteenth centuries were not without apocalyptic prophets and movements. The Nuqtavi resurgence during the last half of the sixteenth century – proclaiming the arrival of the Mahdi and the imminence of the Grand Conjunction – was especially important.

The Mughal empire, in contrast to the Safavid and Ottoman, was larger, wealthier, and more diverse. In India, furthermore, Muslims were a minority – both among the ruling class and in the empire at large. The Mughal calendar borrowed the Indic definitions of the day, and the Mughal ruling elite employed Indic as well as Islamic astrologers. The syncretic nature of the Mughal approach can be seen in the observatories built by Rajah Jai Singh for the Mughal emperor Muhammad Shah. The Shahjahanabad Jantar Mantar, an enlarged replica of Ulugh Beg's institution at Samarqand, was headed by a Hindu astronomer/astrologer who knew the Islamic scientific literature as well as the Indic and whose task was to update the *Zij-i Ilkhani*.

In Mughal India, the ceremonial round was also more complex. An amalgam of the Islamic, the Indic, and the invented, the ritual calendar that Akbar constructed included the four festivals of the early Umayyads, Nau Ruz of the Safavids, the major Hindu festivals, and the Imperial Birthday Festival. All, except for the Islamic, were challenged by the conservative ulama: Nau Ruz for its Zoroastrian ancestry, the Hindu festivals for their idolatry, and the Imperial Birthday Ceremony for its hubris and its Indic inspiration.

Mughal chronology included an invented element as well. To handle the fiscal challenge posed by the Hijra Era, the Mughals first introduced a new taxation era (the Fasli), which, like the Kharaji and the Maliye, had to be periodically adjusted. Soon after, however, they adopted the Turkish Animal Era for fiscal and accounting purposes while creating the Divine Era for histories and public documents. Like the Imperial Birthday Ceremony, the Tarikh-i Ilahi was judged un-Islamic, drawing the ire of the traditional ulama.

The millenarian prophets and movements of sixteenth- and early-seventeenth-century India were also diverse. Two indigenous Sufi orders – the Mahdawiyya (led by Sheikh Mustafa Gujerati) and the Naqshbandi (led by Sheikh Ahmad Sirhindi) – preached revolution and reform to a wide spectrum of the population. The major influence on the Mughal court, however, was a Nuqtavi renegade from Safavid Iran. Sharif Amuli trumpeted the apocalyptic predictions of Mahmud Pasikhani and the Mahdist self-claims of Shah Ismail into the receptive ears of Akbar and his court.

The Ottoman Empire, like the Mughal, was populous, powerful, and diverse. But because its ruling class was composed of converts or the descendants of converts and based its claim to authority and legitimacy on its role as successor to the Caliphate and as protector of the Holy Cities, the Ottoman temporal system was the closest of the three to the lunar Hijra model of early Umayyads. Because of their devotion to the fine points of Islamic ritual, the Ottomans developed an intense interest in timekeeping: developing ever more accurate techniques and instruments to chart the movements of the heavenly bodies and to determine the times of prayer and the beginning and end of the fast days and the fast month. In the sixteenth century, their *muneccims* founded a state-of-the-art observatory and built some of the most accurate mechanical time-pieces in the Eurasian world.

The Ottoman ceremonial round remained the closest of the three to the ritual cycle of the early community. The Ottomans dropped Ashura (the signature celebration of the hated Safavids) and added two new ceremonies. But the send-off of the Hajj caravan and the Imperial Circumcision Festival were more modifications of traditional observances than they were wholesale creations and neither aroused any opposition.

The Ottoman commitment to traditional chronology, however, made developing a workable system for collecting agrarian taxes and paying military salaries extremely difficult. Their unwillingness to adopt an independent solar era for administrative purposes left the Ottomans, like the Umayyads, with the difficult problem of adjusting fiscal and lunar eras – *sivi*s or skipping for the Ottomans and *izdilaq* or sliding for the Umayyads. By contrast, the exclusive use of the Hijra Era by Ottoman historians and administrators relieved them of the dating problems of both the Safavids (finding accurate equivalents across eras) and the Mughals (conservative opposition to new eras).

For the Ottomans the mid to late sixteenth century was also a time of millenarian ferment. From the reign of Sultan Suleiman in the mid-sixteenth century to the reign of Murad III at the end of the century apocalyptic ideas (especially the use of the epithet *sahib kiran*, based on the conjunction astrology of Abu Mashar), became increasingly popular. Because of the popularity of the Hijra Era, the end of the millennium played a more prominent role in Ottoman apocalypticism than it did for either the Mughals or the Safavids.

Time – how to define and organize it – was an important issue for Muhammad and the fledgling Muslim community. The cultic and ceremonial parameters of the new religion gave the early Islamic astronomers/astrologers an intense preoccupation with the calendrical, ceremonial, and chronological aspects of time. As they grappled with the demands of the new dispensation, these men forged the Greek, Indian, and Iranian elements that they had inherited into a new temporal synthesis, along the way advancing the observational and mathematical frontiers of astronomy while producing astrological speculations of ingenuity and plausibility. With this common heritage as a starting point, the *munajjims* of the Safavid, Mughal, and Ottoman empires created a new set of temporal theories. These new concepts were doubly revelatory. On the one hand, they highlighted the truly fundamental practices of the new religion (Ramadan fast, daily prayers, Meccan pilgrimage, lunar calendar and era). On the other, they offered examples of its accommodative genius – illustrating the ways in which Umayyad ideas were blended with Indic, Iranian, and Mediterranean, arriving in each case at a combination that was true both to the deepest impulses of the revelation as well as to the unique spirit of the indigenous culture. In the process of working through the problems associated with the new temporal concepts, Islamic astronomers/astrologers pushed their way to the forefront of their field. Until approximately 1600, leadership in the theory and practice of both astronomy and astrology rested in the Islamic, not in the European, world.

Select Bibliography

Abu Ma'sar on Historical Astrology, ed. and trans. Keiji Yamamoto and Charles Burnett. 2 vols. Leiden: Brill, 2000.

Alderson, A. D. *The Structure of the Ottoman Dynasty*. Oxford: Clarendon Press, 1956.

Allami, Abu al-Fazl. *The A'in-i Akbari*, ed. D. C. Phillott, trans. H. Blochmann. 3 vols. Calcutta: Bibliotheca Indica, 1927–1949; reprint ed. Delhi: Low Price Publications, 1989.

Akbar Nama, trans. H. Beveridge. 3 vols. Calcutta: Bibliotheca Indica, 1948.

Alvi, M. A. and A. Rahman. *Fathullah Shirazi: A Sixteenth Century Indian Scientist*. New Delhi: National Institute of Sciences of India, 1968.

And, Metin. *A History of Theater and Popular Entertainment in Turkey*. Ankara: Forum, 1964.

Ansar, S. M. Razallah, "On the Transmission of Islamic Astronomy to Medieval India," in *The Tradition of Astronomy in India: Jyotihsastra*, B. V. Subbarayappa, ed., 345–73. New Delhi: Centre for Studies in Civilizations.

Aston, Margaret. "The Fiery Trigon Conjunction: An Elizabethan Astrological Prediction." *ISIS* (1970): 159–87.

Atil, Esin Aka. "Surname-I Vehbi: An Eighteenth Century Ottoman Book of Festivals." Ph.D. diss., University of Michigan, 1969.

"The Story of an Eighteenth-Century Ottoman Festival." *Muqarnas* 10 (1993): 181–200.

Aveni, Anthony. *Empires of Time: Calendar, Clocks, and Cultures*. New York: Basic Books, 1989.

Babayan, Kathryn. *Mystics, Monarchs, and Messiahs: Cultural Landscapes of Early Modern Iran*. Cambridge, MA: Harvard University Press, 2003.

al-Badaoni, Abd al-Qadir. *Muntakhabu-T-Tawarikh*, translated by George S. A. Ranking. 3 vols. Reprint ed. Delhi: Darah-I Adabiyat-I Delli, 1973.

Bagci, Serpi and Massumeh Farhad, eds. *Falnama: The Book of Omens*. London: Thames and Hudson, 2010.

al-Biruni, Abu Rayhan. *Chronology of Ancient Nations*, trans. C. Eduard Sachau. London: Oriental Translations Fund, 1879.

Blake, Stephen P. *Half the World: The Social Architecture of Safavid Isfahan, 1590–1722*. Costa Mesa, CA: Mazda Publishers, 1999.

Shahjahanabad: The Sovereign City in Mughal India, 1639–1739. Cambridge: Cambridge University Press, 1991.

Bouhdiba, Abdelwahab. *Sexuality in Islam*. London: Routledge, 2007.

Campion, Nicholas. *The Great Year: Astrology, Millenarianism, and History in the Western Tradition*. London: Arkana, 1994.

Chardin, Jean. *Voyages du Chevalier Chardin, en Perse ...* 10 vols. Paris: Le Normant, 1811.

Combs-Schilling, M. E. *Sacred Performances: Islam, Sexuality, and Sacrifice*. New York: Columbia University Press, 1989.

Daftary, Farhad. *A Short History of the Ismailis*. Edinburgh: Edinburgh University Press, 1998.

Darling, L. T. *Revenue-Raising and Legitimation: Tax Collection and Financial Administration in the Ottoman Empire, 1560–1660*. Leiden: E. J. Brill, 1997.

Dershowitz, Nachum and Edward M. Reingold. *Calendrical Calculations*. Cambridge: Cambridge University Press, 1997.

Duncan, David Ewing. *Calendar: Humanity's Epic Struggle to Determine a True and Accurate Year*. New York: Avon Books, 1998.

Faroqhi, Suraiya. *Subjects of the Sultan: Culture and Daily Life in the Ottoman Empire*. London: I. B. Tauris, 2000.

The Cambridge History of Turkey, vol. 3: The Later Ottoman Empire, 1603–1839. Cambridge: Cambridge University Press, 2006.

Faroqhi, Suraiya N. and Kate Fleet, eds. *The Cambridge History of Turkey, vol. 2: The Ottoman Empire as a World Power, 1453–1603*. Cambridge: Cambridge University Press, 2012.

Friedman, Y. *Sheikh Ahmad Sirhindi: An Outline of His Thought and a Study of His Image in the Eyes of Posterity*. Montreal: McGill University Press, 1971.

Fleischer, Cornell. *Bureaucrat and Intellectual in the Ottoman Empire: The Historian Mustafa Ali (1541–1600)*. Princeton, NJ: Princeton University Press, 1986.

"Ancient Wisdom and New Sciences: Prophecies at the Ottoman Court in the Fifteenth and Early Sixteenth Centuries," in Serpi, Bagci, and Massumeh Farhad, eds., 231–44. *Falnama: The Book of Omens*. London: Thames and Hudson, 2010.

"The Lawgiver as the Messiah: The Making of the Imperial Image in the Reign of Suleyman." in *Soliman le magnifiqe et son temps*, ed. Gilles Veinstein, 159–77. Paris: La Documentation françoise, 1992.

"Seer To Sultan: Haydar-i Remmal and Sultan Suleyman," in *Cultural Horizons: A Festschrift in Honor of Talat S. Halman*, Jayne L. Warner, ed., 290–9. Syracuse: Syracuse University Press, 2001.

"Shadows and Shadows: Prophecy in Politics in 1530s Istanbul." *International Journal of Turkish Studies* 13 (2007): 51–62.

Ghori, S. A. Khan. "Development of Zij Literature in India," in *The Tradition of Astronomy in India: Jyotihsastra*, B. V. Subbarayappa, ed. 383–415. New Delhi: Centre for Studies in Civilizations, 2008.

Gollaher, David. *Circumcision: A History of the World's Most Controversial Surgery*. New York: Basic Books, 2000.

Grafton, Anthony. *Joseph Scalinger: A Study in the History of Classical Scholarship*. 2 vols. Oxford: Clarendon Press, 1983–94.
Hunsberger, Alice C. *Nasir Khusraw, the Ruby of Badakshan*. New York: I.B. Tauris, 2000.
Ihsanoglu, E. *Science, Technology, and Learning in the Ottoman Empire*. London: Ashgate Publishing House, 2004.
Inalcik, Halil. *The Ottoman Empire: Classical Age, 1300–1600*. London: Weidenfeld, 1973.
Imber, Colin. *The Ottoman Empire, 1300–1650: The Structure of Power*. London: Palgrave Macmillan, 2002.
Islam, Riazul. *Calendar of Persian Documents (1500–1700)*. 2 vols. Karachi: Iran Cultural Foundation, 1979.
Issawi, Charles. "The Ottoman Empire in the European Economy." *Journal of the Economic and Social History of the Orient* 1 (1957): 109–17.
Jackson, Peter and Laurence Lockhart, eds. *The Cambridge History of Iran, vol 6: The Timurid and Safavid Periods*. Cambridge: Cambridge University Press, 1986.
Jahangir, Nur al-Din Salim. *The Tuzuk-I Jahangiri*, ed. Henry Beveridge, trans. Alexander Rogers. 2 vols. Delhi: Munshiram Manoharlal, 1968.
Kennedy, E. S. *Studies in the Islamic Exact Sciences*. Beirut: American University of Beirut, 1983.
Kennedy, E. S. and David Pingree. *The Astrological History of Masha`allah*. Cambridge, MA: Harvard University Press, 1971.
Lewis, Bernard and P. M. Holt, eds. *Historians of the Middle East*. London: Oxford University Press, 1962.
Lewis, Raphaela. *Everyday Life in Ottoman Turkey*. London: Batsford, 1971.
Lipton G. A. "Muhibb Allah Ilahabadi's Taswiya Contextualized," in *Muslim Cultures in the Indo-Iranian World During the Early-Modern and Modern Periods*, Denis Herman, and Fabrizio Speziale, eds., 475–98. Berlin: Klaus Schwarz Verlag, 2010.
Matthee, Rudolph. *The Politics of Trade in Safavid Iran: Silk for Silver, 1600–1730*. Cambridge: Cambridge University Press, 1999.
McChesney, R. D. "Four Sources on Shah Abbas's Building of Isfahan." *Muqaranas* 5 (1988), 104–05.
Melville, Charles ed. *Safavid Persia: The History and Politics of an Islamic Society*. London: I. B. Tauris, 1996.
"The Chinese Uighur Animal Calendar in Persian Historiography of the Mongol Period." *Iran* 32 (1994): 83–98.
Moin, Ahmed Azfar. "The Islamic Millennium in Mughal India: An Historiographical Analysis." M.A. thesis, University of Texas at Austin, 2005.
Nasr, Seyyed Hossein. *An Introduction to Islamic Cosmological Doctrines*. Cambridge, MA: Harvard University Press, 1964.
Islamic Philosophy From its Origin to the Present. Albany: State University of New York Press, 2006.
Newman, Andrew. *Safavid Iran: Rebirth of a Persian Empire*. London: I. B. Tauris, 2008.
Newman, Andrew, ed. *Society and Culture in the Early Modern Middle East, Studies on Iran in the Safavid Period*. Leiden: Brill, 2003.

North, John David and Roy Porter. *The Norton History of Astronomy and Cosmology*. New York: Norton, 1994.

Pierce, Leslie. *The Imperial Harem: Women and Sovereignty in the Ottoman Empire*. New York: Oxford University Press, 1993.

Pingree, David. *The Thousands of Abu Ma`shar*. London. The Warburg Institute, 1968.

Quinn, Sholeh. *Historical Writing During the Reign of Shah Abbas: Ideology and Legitimacy in the Safavid Chronicles*. Salt Lake City: University of Utah Press, 2000.

Rahimi, Babak. "The Rebound Theater State: The Politics of the Safavid Camel Sacrifice Rituals, 1598–1695." *Iranian Studies* (2004) 37: 451–78.

Richards, E. G. *Mapping Time: The Calendar and Its History*. Oxford: Oxford University Press, 1999.

Richards, John. *The Mughal Empire*. Cambridge: Cambridge University Press, 1993.

Rizvi, Sajjad H. "Mir Damad and the Debate on Hudut-i Dahri in India," in Denis Herman, and Fabrizio Speziale, eds., 449–74. *Muslim Cultures in the Indo-Iranian World During the Early-Modern and Modern Periods*. Berlin: Klaus Schwarz Verlag, 2010.

Rizvi, S. A. A. *The Religious and Intellectual History of the Muslims in Akbar's Reign*. Delhi: Munshiram Manoharlal, 1975.

The Muslim Revivalist Movements in Northern India in the Sixteenth and Seventeenth Centuries. Agra: Agra University, 1965.

The Socio-Intellectual History of Isna' Ahsari Shi'is in India. 2 vols. Canberra: Ma'rifat Publishing, 1986.

Robinson, Francis. "Ottomans-Safavids-Mughals: Shared Knowledge and Connective Systems." *Journal of Islamic Studies* 8 (1997): 158–9.

Ross, E. Dennison. *An Alphabetical List of the Feasts and Holidays of the Hindus and Muhammadans*. New Delhi: Imperial Record Department, 1914.

Sahillioglu, H. "Sivis Year Crises in the Ottoman Empire," in *Studies in the Economic History of the Middle East*, M. A. Cook, ed., 230–54. Oxford: Oxford University Press, 1970.

Saliba, George. "The Role of the Astrologer in Medieval Islamic Society." *Bulletin d'Etudes Orientales* 44 (1992): 45–68.

Savory, Roger. *Iran under the Safavids*. Cambridge: Cambridge University Press, 1980.

Sayili, Aydin. *The Observatory in Islam*. Ankara: Turk Tarih Kumumu Basimevi, 1988.

Shaw, Stanford. *History of Ottoman Empire. 2 vols.* Cambridge: Cambridge University Press, 1967–77.

Siddiqi, Abdus Sattar. "Construction of Clocks and Islamic Civilization." *Islamic Culture* 1 (1927): 245–51.

Siddiqui, I. H. "Nuqtavi Thinkers at Akbar's Court: A Study of Their Impact on Akbar's Religious and Political Ideas." *Islamic Culture* 72 (1998): 65–82.

Srivastava, A. L. *Akbar the Great*. 2 vols. Delhi: Shiva Lal Agarwal and Company, 1962–68.

Stillman, Myra and Beulah Tannebaum. *Understanding Time: The Science of Clocks and Calendars*. New York: Whittlesey House, 1958.

Stout, Robert E. "The Sur-I Humayun of Murad III: A Study of Ottoman Pageantry and Entertainment." Ph.D. diss., Ohio State University, 1966.

Subrahmanyam, Sanjay. "Connected Histories: Notes towards a Reconfiguration of Early Modern Eurasia." *Modern Asian Studies* 31 (1997): 735–62.

"Turning the Stones Over: Sixteenth-Century Millenarianism from the Tagus to the Ganges." *The Indian Economic and Social History Review* 40 (2003): 129–61.

Taqizadeh, H. "Various Eras and Calendars Used in the Countries of Islam." *BSOAS* 9 (1939): 110–12; *BSOAS* 10 (1940): 107–32.

Tekeli, Sevim. *The Clocks in the Ottoman Empire in the 16th Century and Taqi al Din's "The Brightest Stars for the Construction of the Mechanical Clocks."* Ankara: Ankara Universitesi Basimevi, 1966.

Terzioglu, Derin. "The Imperial Circumcision Festival of 1582: An Interpretation." *Muqarnas* 12 (1995): 84–100.

von Grunebaum, G.E. *Muhammadan Festivals*. New York: Henry Schuman, 1951.

Zerubavel, Eviatar. *The Seven Day Circle: The History and Meaning of the Week*. New York: The Free Press, 1985.

Hidden Rhythms: Schedules and Calendars in Social Life. Chicago: University of Chicago Press, 1982.

Index

Note to Index: An *n* after a page numbers denotes a note on that page.